BIOLOGICAL AND MED
BIOMEDICAL ENGINEF

MW00844267

More information about this series at http://www.springer.com/series/3740

BIOLOGICAL AND MEDICAL PHYSICS, BIOMEDICAL ENGINEERING

The fields of biological and medical physics and biomedical engineering are broad, multidisciplinary and dynamic. They lie at the crossroads of frontier research in physics, biology, chemistry, and medicine. The Biological and Medical Physics, Biomedical Engineering Series is intended to be comprehensive, covering a broad range of topics important to the study of the physical, chemical and biological sciences. Its goal is to provide scientists and engineers with textbooks, monographs, and reference works to address the growing need for information.

Books in the series emphasize established and emergent areas of science including molecular, membrane, and mathematical biophysics; photosynthetic energy harvesting and conversion; information processing; physical principles of genetics; sensory communications; automata networks, neural networks, and cellular automata. Equally important will be coverage of applied aspects of biological and medical physics and biomedical engineering such as molecular electronic components and devices, biosensors, medicine, imaging, physical principles of renewable energy production, advanced prostheses, and environmental control and engineering.

Raquel Cruz Conceição
Johan Jacob Mohr • Martin O'Halloran

Editors

An Introduction to Microwave Imaging for Breast Cancer Detection

 Springer

Editors
Raquel Cruz Conceição
Universidade de Lisboa
Lisboa, Portugal

Johan Jacob Mohr
København NV
Denmark

Martin O'Halloran
Electrical and Electronic Engineering
National University of Ireland Galway
Galway, Ireland

ISSN 1618-7210   ISSN 2197-5647   (electronic)
Biological and Medical Physics, Biomedical Engineering
ISBN 978-3-319-80231-2   ISBN 978-3-319-27866-7   (eBook)
DOI 10.1007/978-3-319-27866-7

Printed on acid-free paper

This Springer imprint is published by Springer Nature
The registered company is Springer International Publishing AG Switzerland

Preface

The genesis of this book was a proposal by Dr. Panagiotis Kosmas and Prof. Yifan Chen to co-author a review paper which would provide an overview of the current state of the art in both microwave breast imaging and microwave breast tumour classification techniques. Later, following discussions with Springer, it was decided that this work might be of interest to a broader community of researchers and the Springer book would be a suitable format for such a review document.

Together with Dr. Martin O'Halloran, they wrote an abstract for a Springer book which broadened the scope of the original document to include quantitative and qualitative imaging methods, as well as a review of existing prototype systems. With this in mind, the team of authors was broadened to include Dr. Tonny Rubæk and Dr. Johan Mohr (microwave tomography); Mr Muhammad Adnan Elahi, Dr. Dallan Byrne, Dr. Edward Jones and Dr. Martin Glavin (confocal microwave imaging); and Ms Marggie Jones and Dr. Raquel Conceição (microwave classification). Since then, the editorial team has been composed of Johan Mohr, Raquel Conceição and Martin O'Halloran.

The team would like to express their thanks to Ho Ying Fan, Karin Louzada, Catherine Rice, Tom Spicer and Cindy Zitter at Springer for their continued support in the development of this book. While the coordination of the book with a large team of contributors has been challenging, it has been equally as rewarding.

The editorial team would also like to acknowledge our many colleagues, whose work has contributed to each and every chapter. In particular, we would like to thank the following researchers (listed in alphabetical order):

- Ms. Bárbara Oliveira
- Dr. Brian McGinley
- Dr. Daniel Flores-Tapia
- Mr. Diego Rodriguez-Herrera
- Dr. Fan Yang
- Dr. Fearghal Morgan
- Mr. Hugo Medeiros
- Prof. Stephen Pistorius.

We would like to acknowledge our past and current institutions (listed in alphabetical order):

- Department of Electrical and Electronic Engineering, University of Bristol, United Kingdom
- Department of Electrical Engineering, Technical University of Denmark, Kongens Lyngby, Denmark
- Electrical and Electronic Engineering, National University of Ireland, Galway, Ireland
- Institute of Biomedical Engineering, Department of Engineering Science, University of Oxford, United Kingdom
- Instituto de Biofísica e Engenharia Biomédica, Faculdade de Ciências, Universidade de Lisboa, Lisbon, Portugal
- King's College London, Strand London, United Kingdom
- Medical Device Research Group, Translational Research Facility, National University of Ireland, Galway, Ireland
- Newcastle University, Newcastle upon Tyne, United Kingdom
- South University of Science and Technology of China, Shenzhen, China

Also, we would like to acknowledge the various funding agencies who have supported our work during the completion of this book (listed in alphabetical order):

- EPSRC with grant numbers: EP/J00717X/1 and EP/J007293/1
- Fundação para a Ciência e Tecnologia with postdoctoral grant reference SFRH/BPD/79735/2011 and the strategic plan of the Instituto de Biofísica e Engenharia Biomédica PEst-OE/SAU/UI0645/2011
- Guangdong Natural Science Funds for Distinguished Young Scholar (2013) under grant S2013050014223
- Marie Curie Intra-European Fellowship within the 7th European Community Framework Programme under REA grant agreement number 301269
- Science Foundation Ireland: 11/SIRG/I2120
- South University of Science and Technology of China Internal Grants (2015): FRG-SUSTC1501A-49 and FRG-SUSTC1501A-63

Finally, we would like to thank the MiMed COST Action (TD1301) for supporting all aspects of the development of this book, including three short term scientific missions. The book itself is very much aligned with the goals of the MiMed Action, in terms of sharing experience and expertise, and the development of long-lasting collaborative networks.

Lisbon, Portugal   Raquel Cruz Conceição
Kongens Lyngby, Denmark   Johan Jacob Mohr
Galway, Ireland   Martin O'Halloran
March 2016

Contents

List of Contributors

Dallan Byrne Faculty of Engineering, University of Bristol, Bristol, UK
formerly at the Electrical and Electronic Engineering, National University of Ireland, Galway, Ireland

Department of Electrical & Electronic Engineering, University of Bristol, Bristol, UK

Yifan Chen formerly at Newcastle University, Newcastle upon Tyne, UK

South University of Science and Technology of China, Shenzhen, China

Raquel Cruz Conceição formerly at the Department Engineering Science, Institute of Biomedical Engineering, University of Oxford, Oxford, UK

Faculdade de Ciências, Instituto de Biofísica e Engenharia Biomédica, Universidade de Lisboa, Lisbon, Portugal

Muhammad Adnan Elahi Electrical and Electronic Engineering, National University of Ireland, Galway, Ireland

Martin Glavin Electrical and Electronic Engineering, National University of Ireland, Galway, Ireland

Edward Jones Electrical and Electronic Engineering, National University of Ireland, Galway, Ireland

Marggie Jones Medical Device Research Group, Translational Research Facility, National University of Ireland, Galway, Ireland

Panagiotis Kosmas King's College London, Strand, London, UK

Johan Jacob Mohr formerly at the Department of Electrical Engineering, Technical University of Denmark, Kongens Lyngby, Denmark

Mellanox Technologies, Roskilde, Denmark

Martin O'Halloran Electrical and Electronic Engineering, National University of Ireland, Galway, Ireland

Tonny Rubæk formerly at the Department of Electrical Engineering, Technical University of Denmark, Kongens Lyngby, Denmark

OHB Systems, Bremen, Germany

Acronyms

2-D	Two Dimensional
3-D	Three Dimensional
A/D	Analog-to-Digital
ABS	Acrylonitrile Butadiene Styrene
ANNs	Artificial Neural Networks
ART	Algebraic Reconstruction Technique
BIM	Born Iterative Method
CDF	Cumulative Distribution Function
CGLS	Conjugate Gradient Least Square
CMI	Confocal Microwave Imaging
CNR	Complex Natural Resonance
CS	Coarse-Shape
CSCS	Coarse-Size-Coarse-Shape
CSFS	Coarse-Size-Fine-Shape
CSI	Contrast Source Inversion
CF	Coherency Factor
CT	Computer X-ray Tomography
CW	Continuous Wave
DAS	Delay-and-Sum
DBIM	Distorted Born Iterative Method
DCIS	Ductal Carcinoma In Situ
DDA	Discrete Dipole Approximation
DMAS	Delay-Multiply-and-Sum
DORT	Decomposition of the Time Reversal Operator
DWT	Discrete Wavelet Transform
EIS	Electrical Impedance Spectroscopy
FD	Frequency Domain
FDTD	Finite-Difference Time-Domain
FEM	Feature Extraction Methods
FFT	Fast Fourier Transform

FIR	Finite-Impulse Response
FS	Fine-Shape
FSCS	Fine-Size-Coarse-Shape
FSFS	Fine-Size-Fine-Shape
FWHM	Full-Width Half-Maximum
GA	Genetic Algorithms
GRS	Gaussian Random Sphere
ICA	Independent Component Analysis
IF	Intermediate Frequency
IDAS	Improved Delay and Sum
IDC	Invasive Ductal Carcinoma
IFFT	Inverse Fast Fourier Transform
ILC	Invasive Lobular Carcinoma
IMAS	Improved-Multiply-and-Sum
K	Kernel
LCIS	Lobular Carcinoma In Situ
LDA	Linear Discriminant Analysis
LDB	Local Discriminant Basis
LLR	Log-Likelihood Ratio
LSQR	Least Square QR (Factorisation)
MAMI	Multistatic Adaptive Microwave Imaging
MDM	Multistatic Data Matrix
MIMO	Multiple-Input Multiple-Output
MIS	Microwave Impedance Spectroscopy
MISP	Microlobulated/Spiculated
MIST	Microwave Imaging Space-Time
MMHP	Modified Hermite Polynomial
MR-CIS	Multiple Regularised Contrast Source Inversion
MRI	Magnetic Resonance Imaging
MST	Modulated Scattering Technique
MWI	Microwave Imaging
NEAT	Neuroevolution Through Augmenting Topologies
NP	Neyman-Pearson
OMA	Oval/Macrolobulated
OUT	Object Under Test
PCA	Principal Component Analysis
PDF	Probability Density Function
PEC	Perfect Electric Conductor
PET	Positron Emission Tomography
PSO	Particle Swarm Optimisation
QDA	Quadratic Discriminant Analysis
QF	Quality Factor
RBF	Radial Basis Function
RCB	Robust Capon Beamformer
ROC	Receiver Operating Characteristic

ROEEC	Receiver Outage Error Exponent Characteristics
RTS	Radar Target Signature
Rx	Receive Antenna
SBR	Synthesised Broadband Reflector
SC	Selection Combining
SMR	Signal-to-Mean Ratio
SMXR	Signal-to-Max Ratio
SNN	Spiking Neural Network
SOM	Self-Organising Map
SWCNT	Single-Walled Carbon Nanotubes
SP6T	Single-Pole 6 Through
SVD	Singular Value Decomposition
SVM	Support Vector Machines
SWCNTs	Single-Walled Carbon Nanotubes
TD	Time Domain
TE	Transverse Electric
TEM	Transverse Electromagnetic
TF/SF	Total-Field/Scattered-Field
TG-RCB	Transmitter-Grouping Robust Capon Beamformer
TLS	Total Least Square
TRO	Time Reversal Operator
Tx	Transmit Antenna
UPML	Uniaxial Perfectly Matched Layer
UWB	Ultrawide Band
VNA	Vector Network Analyser

Chapter 1
Introduction

Johan Jacob Mohr, Martin O'Halloran, and Raquel Cruz Conceição

Over the last two decades, microwave imaging (MWI) has been investigated as a novel imaging and diagnostics technique for detection of breast cancer. A number of early small-scale clinical experiments have clearly illustrated the potential of the technology, while also revealing some significant remaining challenges. These technical challenges must be addressed before MWI is accepted as a viable alternative (or complement) to other medical imaging techniques such as X-ray, ultra-sound, and magnetic resonance imaging.

The purpose of this book is two-fold: firstly, we wish to present the background and theory of the most basic MWI and diagnostic techniques at a level which is accessible to graduate-level newcomers to the field; secondly, we wish to give an overview of the most recent and relevant developments in the field of MWI. In this way, we hope to bridge the gap between more easily accessible overviews of MWI, e.g. [2–5, 7], which presents broad overview of results and techniques, but

J.J. Mohr (✉)
formerly at the Department of Electrical Engineering, Technical University of Denmark, Kongens Lyngby, Denmark

Mellanox Technologies, Roskilde, Denmark
e-mail: johan_jacob_mohr@yahoo.dk

M. O'Halloran
Electrical and Electronic Engineering, National University of Ireland, Galway, Ireland
e-mail: martin.ohalloran@nuigalway.ie

R.C. Conceição
Faculdade de Ciências, Instituto de Biofísica e Engenharia Biomédica, Universidade de Lisboa, Lisbon, Portugal
e-mail: raquelcruzconceicao@gmail.com

© Springer International Publishing Switzerland 2016   1
R.C. Conceição et al. (eds.), *An Introduction to Microwave Imaging for Breast Cancer Detection*, Biological and Medical Physics, Biomedical Engineering, DOI 10.1007/978-3-319-27866-7_1

lacks the detail required to gain a complete understanding of the systems, and the very detailed texts such as [1, 6] which attempt to include a complete treatise of the theory, but can appear technically daunting to novices in the field. It should be noted that MWI for breast cancer detection is a growing field of research, and as such many new and promising research groups are continually emerging. As such, this book can only present work from the established leading research groups at the time of publishing.

The book begins by describing *breast anatomy* and cancer development. Both of these topics are addressed from the perspective of MWI. Next, a summary of both historical and more recent studies of the *dielectric properties* of breast tissue is presented, and how these studies contribute to the current understanding of the microwave breast imaging problem is discussed.

Chapter 3 examines the underlying principles of *Microwave Tomography*, an imaging technique where the coarse dielectric profile of the breast is reconstructed. In this chapter, the scattering problem is presented, in the form of integral equations, and methods for solving the non-linear imaging problem are described.

Chapter 4 is focused on *Confocal Microwave Imaging* or radar-based imaging. Rather than seeking to reconstruct the dielectric profile of the breast, radar-based methods aim to identify the presence and location of significant dielectric scatterers within the breast. In this chapter, a representative range of algorithms are described and evaluated on simulated data-sets, and general trends are drawn in the context of different levels of dielectric heterogeneity. The emerging idea of contrast enhanced imaging is also considered.

Apart from using reflected microwave energy to reconstruct images of the breast, the tumour reflections (or radar target signatures) may contain additional information on the shape and size of the tumour. This information could potentially be used for tumour classification. In Chap. 5, *Tumour Classification*, different numerical models are first presented, which can be used to realistically mimic the size, shape, and growth patterns of breast tumours in electromagnetic numerical models. Next, a range of tumour classification algorithms based on the radar target signatures of tumours is presented.

Finally, the book concludes with a snap-shot of current *Experimental Systems*. For clarity, these systems have been categorised into three groups: operational, tomographic, and radar-based. The main characteristics of each system are presented and how they have been evaluated is discussed. Importantly, references to the most recent studies are included, rather than the earliest, which are often found to be obsolete.

Overall, this book provides a comprehensive but accessible map of the microwave imaging and diagnostic research, and will act as a useful introductory guide for novice researchers in this promising field of medical imaging.

References

[1] Chew W (1995) Waves and fields in inhomogeneous media. IEEE Press, New York
[2] Fear E (2005) Microwave imaging of the breast. Techn in Cancer Research and Treatment 4(1):69–82
[3] Fear EC, Hagness SC, Meaney PM, Okoniewski M, Stuchly MA (2002) Enhancing breast tumor detection with near-field imaging. IEEE Microwave Magazine 3(1):48–56
[4] Hassan AM, El-Shenawee M (2011) Review of electromagnetic techniques for breast cancer detection. IEEE Reviews in Biomedical Engineering 4:103–118, DOI 10.1109/RBME.2011.2169780
[5] Nikolova N (2011) Microwave imaging for breast cancer. IEEE Microwave Magazine 12(7):78–94, DOI 10.1109/MMM.2011.942702
[6] Pastorino M (2010) Microwave Imaging. Wiley, Hoboken, NJ
[7] Paulsen KD, Meaney PM, Gilman LC (2005) Alternative breast imaging four model-based approaches. Springer, New York, NY

Chapter 2
Anatomy and Dielectric Properties of the Breast and Breast Cancer

Martin O'Halloran, Dallan Byrne, Raquel Cruz Conceição, Edward Jones, and Martin Glavin

In this chapter, the anatomy of the breast is presented first. The anatomy of the breast is described primarily from the perspective of microwave imaging, with a focus on the various layers and structure of the breast with differing dielectric properties. Next, the disease of breast cancer is discussed, including the cellular origins of the disease, the variants or types of the disease and the grading of breast cancer. The chapter concludes with an in-depth review of both historical and recent studies of the dielectric properties of healthy and cancerous breast tissue.

M. O'Halloran (✉) • E. Jones • M. Glavin
Electrical and Electronic Engineering, National University of Ireland, Galway, Ireland
e-mail: martin.ohalloran@nuigalway.ie; edward.jones@nuigalway.ie; martin.glavin@nuigalway.ie

D. Byrne
Faculty of Engineering, University of Bristol, Bristol, UK

formerly at Electrical and Electronic Engineering, National University of Ireland, Galway, Ireland

Department of Electrical & Electronic Engineering, University of Bristol, Bristol, UK
e-mail: dallan.byrne@bristol.ac.uk

R.C. Conceição
Faculdade de Ciências, Instituto de Biofísica e Engenharia Biomédica, Universidade de Lisboa, Lisbon, Portugal
e-mail: raquelcruzconceicao@gmail.com

© Springer International Publishing Switzerland 2016
R.C. Conceição et al. (eds.), *An Introduction to Microwave Imaging for Breast Cancer Detection*, Biological and Medical Physics, Biomedical Engineering, DOI 10.1007/978-3-319-27866-7_2

2.1 Anatomy of the Breast

The shape and size of the breast, and the properties and distribution of the constituent tissues, will significantly influence the design of a microwave breast imaging system [7, 26]. The anatomical transformation of the breast during the lifetime of the woman from puberty to menopause and beyond must also be considered. Several general anatomy texts provide very comprehensive examinations of the development and structure of the normal healthy breast [9, 10], and a brief summary is presented here.

At maturity, the breast has a protruding conical form with a circular base (with a typical diameter of between 7 and 8 cm). The breast consists of between 15 and 20 pyramid-shaped lobes of glandular tissue arranged in a wheel-spoke fashion emanating from the nipple. The nipple is surrounded by a ring of pigmented skin called the areola, which is usually 2.5 cm in diameter. Each lobe is composed of approximately between 20 and 40 lobules that can be subdivided further into functional units known as acini. The acini of the breast are small glands where milk is produced in response to normal hormonal changes. Adipose (fatty) tissue fills the gaps between the lobes and extends throughout the breast, giving the breast its size and shape. Fibrous connective tissue forms a supporting and stabilising framework for the lobes, and is often referred to as Cooper's ligament. Finally, the retromammary fat separates the breast from the major pectoral muscle.

A non-lactating breast normally weighs between 150 and 225 g, with a lactating breast increasing in weight to 500 g. The mean volume of the breast is approximately 285 ml. The breast of a nulliparous woman (a woman who has never given birth) typically has a hemispherical shape with a noticeable flattening above the nipple. The multiparous breast (belonging to a woman who has given birth two or more times) is usually larger and more pendulous, due to the normal hormonal stimulation experienced during pregnancy and lactation.

The breasts of younger women are composed primarily of dense glandular tissue, with a small percentage of the overall breast volume being fat. Therefore, they tend to be more firm and dense than the breasts of older women. In the fortnight before menstruation, the breast ducts enlarge as the overall breast changes in size and consistency. This change is in response to increased levels of the hormone prolactin. During pregnancy (again prompted by hormonal changes), the blood supply to the breasts increases and significant duct, lobular and alveolar growth occurs. As the pregnancy progresses, the breasts tend to enlarge with dilation of the superficial veins. Finally, with the loss of oestrogen during the menopause, a significant change in the consistency of the breast occurs where the lobes of the breast are replaced by fat, and the overall breasts becomes less dense and less firm.

From the perspective of microwave imaging, the anatomy of the breast can be simplified to the following:

• An adipose layer directly below the skin. This layer consists of vesicular cells filled with fat which are collected into lobules and then separated by Cooper's ligament;

- The innermost tissue of the breast consists of mammary glands (lobules that produce milk). Each breast has about 15–20 sections termed lobes with many smaller sections of mammary glands, which are arranged in a circular fashion. These lobes and ducts are also surrounded by 'Cooper's ligament', which has the function of maintaining the inner structure of the breast and supporting the tissue attached to the chest wall;
- Posterior to the breast is the major pectoral muscle, and the 2nd to 6th rib.

Breast tumours typically originate in glandular tissue. The formation of these tumours will be discussed in detail in the next section.

2.2 Breast Cancer

Breast cancer is most common in older women [17] potentially due to the fact that cells have to undergo multiple genetic alterations before becoming malignant [15]. There is also a higher incidence of breast cancer when a family history of the disease exists, or when the patient previously developed breast or any other type of cancer [17].

Breast cancer is generally categorised as either invasive or in situ (also called non-invasive). Invasive cancers are those in which there is a dissemination of cancer cells outside the basement membrane of the ducts and lobules into the surrounding healthy tissue. Conversely, in situ cancers are those in which cancer cells remain within the basement membrane of the lobules and the draining lactiferous ducts [8].

Some of the most frequently occurring breast cancers are as follows:

- Invasive ductal carcinoma (IDC) is the most common type of breast cancer (70–80 % of breast cancer cases) and occurs in the cells that line the ducts of the breasts;
- Invasive lobular carcinoma (ILC) represents about 10 % of breast cancer cases and occurs in the cells that line the lobules of the breast;
- The ductal carcinoma in situ (DCIS) is a type of cancer in which cancerous cells are present inside the ducts, but have not spread to other regions of the breast or body;
- The lobular carcinoma in situ (LCIS) is not a type of cancer *per se*, but the presence of this disease suggests an increased risk of developing cancer.

In general, breast cancer is defined as a growth of undifferentiated (or unspecialised) cells which form a lump. Usually the immune system is capable of destroying the undifferentiated cells which can lead to the formation of a tumour through a process called apoptosis (cell self-destruction). However, if too many mutations occur in cells at the same time, the immune system may not be able to respond appropriately, and masses of tumour cells will be formed [17].

The manner in which proliferation of tumour cells occurs may indicate whether a tumour is benign or malignant. For benign tumours, the growth is controlled and may only be of concern if nearby organs are compressed or if the tumour releases unwanted hormones. Conversely, for malignant tumours the growth is uncontrolled due to a high rate of replication. Malignant cancer usually spreads to other parts of the body by metastases and destroys surrounding healthy tissues [17].

The grade of malignancy of the tumour can be determined by pathologically analysing how premature the cells are within the tumour. The less mature the tumour cells, the older and more widespread the malignant tumour is likely to be, and therefore the lower the chances of successful treatment. The different grades of development at which cells can be found are referred to as differentiation [17]. The cytoskeleton of tumour cells becomes disorganised due to the decrease and disorganisation of microfilaments and microtubules [3], causing the original shape of the cell to be lost (becoming more round) and the process of mitosis (cell replication) to become chaotic leading to both an uncontrolled growth of tissues and a loss of genetic information. Because the surface of the cells changes, the membrane permeability is altered and the regular osmosis process is also affected, causing the tumour tissues to retain more fluid than normal cells. This extra fluid, in the form of bound water, changes the dielectric properties of the tissue.

Additionally, cancer cells are not contact-inhibited, which means that huge masses of cancer cells grow over each other forming multiple layers, and are able to coexist in high concentrations. Furthermore, due to the large growth of cells in tumours, particularly in malignant tumours, networks of capillaries are created in order to nourish the newly formed cells [2]—a process called neoangiogenesis. In [23], it is noted that neoangiogenesis, i.e. growth of new blood vessels, is induced by tumours with a dimension of at least 3 mm. As the size of tumours becomes larger, these networks of capillaries may develop into small veins and arteries which will connect to major blood supply vessels [2].

The increased volume of water within cancerous tissue is responsible for the strong electromagnetic scattering associated with microwave imaging. The increase of sodium and water, particularly in bound water, within the tumour cell leads to greater conductivity and permittivity of tumour tissues [18, 27, 29]. Another feature that may help detect the presence of malignant tumours is the existence of calcifications. However these are only formed when severe necrosis has occurred, i.e. disorderly apoptosis, resulting in groups of dead cells which are not naturally absorbed by the organism [29].

Finally, some additional characteristics inherent to benign and malignant tumours have potential to be helpful in tumour classification. These characteristics include tumour shape, margins, surface texture, depth, localisation and packing density [2, 20, 23, 28]. A subset of these characteristics may be useful for tumour classification using microwave imaging: tumour shape and surface texture. Malignant tumours usually present the following characteristics:

- Irregular, ill-defined and asymmetric shapes;
- Blurred boundaries (lack of sharpness);
- Rough and complex surfaces with spicules or microlobules;
- Non-uniform permittivity variations;

- Distortion in the architecture of the breast;
- Irregular increase of tissue density (due to masses and calcifications).

Conversely, benign tumours tend to have the following characteristics:

- Spherical, oval or at least present well-circumscribed contours;
- Compact;
- Smooth surfaces [1, 2, 16, 25, 28].

In the next section, the dielectric properties of both healthy and cancerous tissue at microwave frequencies are described in detail.

2.2.1 Historical Dielectric Studies

Several historical studies have examined the dielectric properties of the breast and breast cancer. These results are summarised here first for completeness, before the results of more recent studies are presented.

In general, the dielectric properties of normal and cancerous tissue of various human organs have long been the subject of various studies [11–14]. However, focussing on the breast in isolation, in 1984, Chaudhary et al. [5] measured the dielectric properties of ex vivo normal and malignant breast tissue between 3 MHz and 3 GHz at 25 °C. The variation of the dielectric properties of normal and malignant tissue with frequency as established by Chaudhary is shown in Fig. 2.1.

Chaudhary concluded that significant differences existed in the dielectric properties of normal and malignant tissues of the human breast, with the greatest permittivity difference occurring at frequencies below 100 MHz.

In 1988, Surowiec et al. [31] measured the dielectric properties of ex vivo infiltrating breast carcinoma and its surrounding tissue at frequencies between

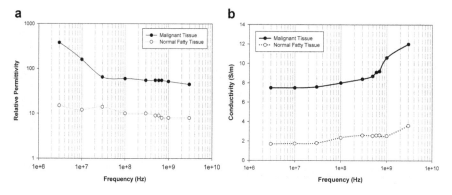

Fig. 2.1 The variation of the relative permittivity (**a**) and conductivity (**b**) of normal and malignant tissue between 3 MHz and 3 GHz as reported by Chaudhary et al. [5]. Images reproduced courtesy of The Electromagnetics Academy

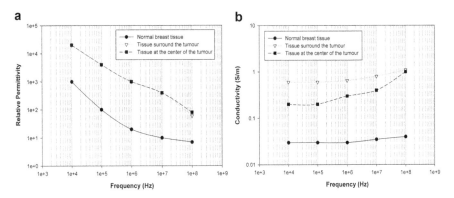

Fig. 2.2 The variation of (**a**) the relative permittivity and (**b**) the conductivity of tumour tissue, the surrounding tissue, and peripheral tissue across the frequency band of 0.02 and 100 MHz as reported by Surowiec et al. [31]. Images reproduced courtesy of The Electromagnetics Academy

20 KHz and 100 MHz. The results were fitted to Cole–Cole dielectric relaxation models [32]. Three categories of tissue were considered by Surowiec et al.:

- The central part of the tumour;
- The tissue immediately surrounding the tumour;
- The peripheral tissue at a distance of approximately 2 cm from the centre of the tumour.

The Cole–Cole models of permittivity and conductivity for the centre of the tumour, the surrounding tissue and peripheral tissue, as reported by Surowiec et al. [31], are shown in Fig. 2.2. Surowiec again observed significantly higher permittivity values for the central part of the tumour and the infiltrating margins compared to tissue taken from the periphery of the tumour. Surowiec suggested that the high permittivity associated with the infiltrating margins of the tumour could be attributed to tumour cell proliferation, and, based on the results the dielectric contrast between normal and malignant tissue, could be potentially used for the detection of breast cancer.

In 1992, Campbell and Land [4] measured the complex permittivity of female breast tissue at 3.2 GHz. Campbell and Land's specific aim was to provide detailed information for microwave thermography applications on the dielectric properties of human female breast tissue at 3.2 GHz. Campbell contended that the dielectric measurements made by Surowiec et al. may have been inaccurate at microwave frequencies due to the fact that the samples were collected and stored in physiological saline and that the results at microwave frequencies could be more representative of the saline than the breast tissue sample itself. In [4], four types of tissue were examined: fatty tissue, normal tissue, benign breast tumour tissue and malignant breast tissue. The results are outlined in Table 2.1. While Campbell and Land [4] noted a significant dielectric contrast between normal (fat tissue and all other breast

Table 2.1 Dielectric properties of female breast tissue at 3.2 GHz as reported by Campbell and Land [4]

Tissue type	Relative permittivity	Conductivity (S/m)	Water content (%)
Fatty tissue	2.8–7.6	0.54–2.9	11–31
Normal tissue	9.8–46	3.7–34	41–76
Benign tissue	15–67	7–49	62–84
Malignant tissue	9–59	2–34	66–79

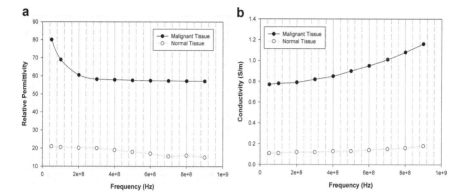

Fig. 2.3 The variation of (**a**) the relative permittivity and (**b**) the conductivity of normal and malignant tissue between 50 and 900 MHz as reported by Joines et al. [19]. Images reproduced courtesy of The Electromagnetics Academy

tissue) and tumour tissue, they also suggested that due to the similarity in dielectric properties of malignant and benign tumours, it might not be possible to distinguish between the two based on dielectric properties alone.

In 1994, Joines et al. [19] measured the dielectric properties of normal and malignant breast tissue from 50 to 900 MHz. The measured permittivity and conductivity are shown in Fig. 2.3. Across the range of tissues examined, Joines observed an average difference in permittivity and conductivity of 233 and 577 %, respectively.

In 2000, Meaney et al. [24] used a clinical prototype of a tomographic microwave imaging system to estimate the dielectric properties of normal breast tissue in vivo. Meaney measured the average permittivity and conductivity of cancer-free breast tissue. The results of this study are shown in Table 2.2. Meaney noted that the average permittivity values of normal tissue at 900 MHz were significantly higher than those previously published in Joines et al.'s ex vivo study [19]. Meaney also suggested that a correlation existed between the average permittivity values and the radiographic density of the tissue, since patients categorised radiographically as having breast tissue with high fat content had an average permittivity value of 31, while patients categorised as having breast tissue with heterogeneously dense tissue had average relative permittivity between 35 and 36.

Table 2.2 Average dielectric properties of female breast tissue at 900 MHz measured in vivo using an active microwave imaging system developed by Meaney et al. [24]

Patient	Age	Average permittivity	Average conductivity (S/m)
1	76	17.22 ± 11.21	0.5892 ± 0.3547
2	57	31.14 ± 4.35	0.6902 ± 0.3650
3	52	36.44 ± 6.24	0.6869 ± 0.3156
4	49	35.43 ± 3.93	0.5943 ± 0.3841
5	48	30.85 ± 7.22	0.6350 ± 0.3550

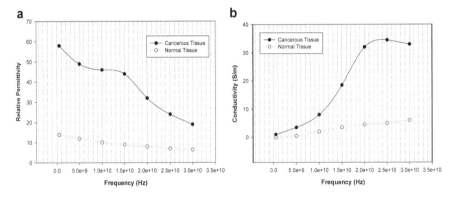

Fig. 2.4 The variation of (**a**) the relative permittivity and (**b**) the conductivity of normal and malignant tissue between 0.5 and 30 GHz as reported by Choi et al. [6]. Images reproduced courtesy of The Electromagnetics Academy

2.3 More Recent Dielectric Studies

In 2004, Choi et al. [6] measured the dielectric properties of breast cancer tissue, along with the properties of metastasised lymph nodes and normal lymph nodes in the frequency band between 0.5 and 30 GHz. The results of the measurements are shown in Fig. 2.4, once again illustrating a dielectric contrast between normal and malignant tissue.

One of the most recent, and arguably most comprehensive examinations of the dielectric properties of normal and malignant tissue was completed by Lazebnik et al. [21, 22] in 2007. Lazebnik sought to compensate for some of the apparent weaknesses of previous studies, such as small patient numbers, the fact that many studies did not exceed 3.2 GHz in frequency, and the limited types of tissues examined. In her first study [21], Lazebnik examined normal breast tissue exclusively. This tissue was obtained from breast reduction surgeries. One of the most significant differences between Lazebnik's first study and previous studies was the histological categorisation of the samples. Each sample under consideration was quantified in terms of the percentage of adipose, glandular and fibroglandular tissue present in the sample.

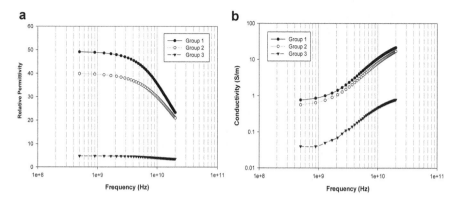

Fig. 2.5 The relative permittivity (**a**) and conductivity (**b**) of normal breast tissue as measured by Lazebnik et al. [21] over the frequency band 0.5–20 GHz. Group 1 represents 0–30 % adipose tissue, group 2 represents 31–84 % adipose and group 3 represents 85–100 % adipose tissue

In order to effectively summarise the data, Lazebnik et al. formed 3 groups of tissue:

- Group 1 contained all samples with 0–30 % adipose tissue (99 samples).
- Group 2 contained all samples with 31–84 % adipose tissue (84 samples).
- Group 3 contained all samples with 85–100 % adipose content (171 samples).

Median permittivity and conductivity curves were created by calculating the fitted values for each sample in the group at 50 equally spaced frequency points. Secondly, the median value at a particular frequency was calculated across all samples within a group. Finally, Cole–Cole equations were used to fit these median values. The Cole–Cole representations for permittivity and conductivity for each tissue group are shown in Fig. 2.5. Lazebnik found significant dielectric heterogeneity in normal breast tissue, as previously suggested by Campbell and Land [4]. Lazebnik suggested that the reason that this level of heterogeneity was not found in previous studies was the location from which samples of normal tissue were taken. In previous studies, the samples of normal tissue were taken from regions distinct from the tumour site, and since tumours typically occur in glandular tissue, these normal samples were likely to have a higher adipose content compared to the glandular tissue surrounding the tumour. Therefore, the dielectric heterogeneity of breast tissue was underestimated. Lazebnik also concluded that the dielectric properties of breast tissue was primarily a function of the adipose content of the tissue. Adjusting for adipose content, Lazebnik found that there only existed a 10 % difference between the conductivity of normal tissue and malignant tissue, and an approximate 8 % difference in permittivity at 5 GHz. Adjusting for adipose and fibroconnective tissue, Lazebnik found no statistical differences between normal glandular and malignant glandular tissues in the breast. This presented a much more difficult imaging scenario than that previously assumed.

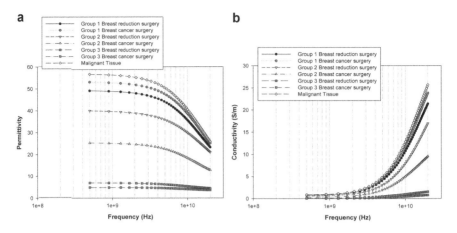

Fig. 2.6 The median relative permittivity (**a**) and conductivity (**b**), (50th percentile) Cole–Cole curves for groups 1, 2 and 3 for normal tissue obtained from reduction surgeries and cancer surgeries. The median relative permittivity (50th percentile) curve of the dielectric properties of samples that contained at least 30 % malignant tissue content is also shown for comparison

The dielectric properties of normal, benign and malignant breast tissues were further addressed in Lazebnik's later study [22]. Once again, Lazebnik attempted to address discrepancies and gaps in previous studies by undertaking a study with a large population size across a very wide frequency band from 0.5 to 20 GHz. The results are summarised in Fig. 2.6.

The measured dielectric values for malignant tissue agreed well with previous studies by Chaudhary [5], Surowiec [30] and Joines [19]. Adjusting for adipose content, Lazebnik et al. found that there only existed a 10 % difference between the conductivity of normal tissue and malignant tissue, and an approximate 8 % difference in permittivity at 5 GHz. Adjusting for adipose and fibroconnective tissue, Lazebnik et al. found no statistical differences between normal glandular and malignant glandular tissues in the breast. Without adjusting for adipose or fibroconnective tissue, Lazebnik et al. observed a 10:1 contrast between normal high adipose tissue and malignant tissue.

Finally, in 2009, Halter et al. presented results from a small-scale clinical study in which estimates of the dielectric properties of malignant breast tissue were obtained using three different methods:

1. Electrical impedance tomography;
2. In vivo measurement using both electrical impedance spectroscopy (EIS) and microwave impedance spectroscopy (MIS) probes;
3. Ex vivo breast cancer specimens, again using EIS and MIS probes.

The importance of this particular study is that it involves in vivo measurements. While the dielectric properties of normal tissue were in general agreement with Lazebnik's studies, Halter observed a significant change in dielectric parameters after tissue excitation and attributed those changes to variations of temperature, tissue dehydration and ischemic effects. Moreover, they also noted that those changes occur within seconds after extraction and may stabilise for hours thereafter. Halter also found that the dielectric properties of in vivo cancerous tissue were significantly higher than those estimated by either electrical impedance tomography or by ex vivo measurements. However, it could be argued that since the dielectric probe was inevitably not sampling the exact same tissue volume in vivo and ex vivo (since the probe had to be removed for tissue excision), it is difficult to draw very definite conclusions from this particular study. Therefore, Lazebnik's studies still remain the main reference study for the estimation of the dielectric properties of healthy and cancerous breast tissue. However, this dielectric heterogeneous breast as presented by Lazebnik suggests quite a challenging imaging scenario from the perspective of microwave imaging.

References

[1] Bindu G, Mathew K (2007) Characterization of benign and malignant breast tissues using 2-d microwave tomographic imaging. Microwave and Optical Technology Letters 49(10):2341–2345
[2] Bridges JE, Hagness S, Taflove A, Popovic M (2002) Microwave discrimination between malignant and benign breast tumors
[3] Cameron I (1981) The transformed cell. Elsevier
[4] Campbell A, Land D (1992) Dielectric properties of female human breast tissue measured in vitro at 3.2 ghz. Physics in medicine and biology 37(1):193
[5] Chaudhary S, Mishra R, Swarup A, Thomas JM (1984) Dielectric properties of normal & malignant human breast tissues at radiowave & microwave frequencies. Indian journal of biochemistry & biophysics 21(1):76
[6] Choi JW, Cho J, Lee Y, Yim J, Kang B, Oh KK, Jung WH, Kim HJ, Cheon C, Lee HD, et al (2004) Microwave detection of metastasized breast cancer cells in the lymph node; potential application for sentinel lymphadenectomy. Breast cancer research and treatment 86(2):107–115
[7] Conceição RC, O'Halloran M, Glavin M, Jones E (2011) Numerical modelling for ultra wideband radar breast cancer detection and classification. Progress In Electromagnetics Research B 34:145–171
[8] Dixon JM (2006) ABC of breast diseases, vol 69. John Wiley & Sons
[9] Drake R, Vogl AW, Mitchell AW (2014) Gray's anatomy for students. Elsevier Health Sciences
[10] Faiz O, Blackburn S, Moffat D (2011) Anatomy at a Glance, vol 66. John Wiley & Sons
[11] Gabriel C (1996) Compilation of the dielectric properties of body tissues at RF and microwave frequencies. Tech. rep., DTIC Document
[12] Gabriel C, Gabriel S, Corthout E (1996a) The dielectric properties of biological tissues: I. literature survey. Physics in medicine and biology 41(11):2231
[13] Gabriel S, Lau R, Gabriel C (1996b) The dielectric properties of biological tissues: Ii. measurements in the frequency range 10 hz to 20 ghz. Physics in medicine and biology 41(11):2251
[14] Gabriel S, Lau R, Gabriel C (1996c) The dielectric properties of biological tissues: III. parametric models for the dielectric spectrum of tissues. Physics in medicine and biology 41(11):2271

[15] Goldblum JR, Weiss SW (2008) Soft Tissue Tumors. Mosby

[16] Guliato D, Rangayyan RM, Carvalho JD, Santiago S, et al (2008) Polygonal modeling of contours of breast tumors with the preservation of spicules. Biomedical Engineering, IEEE Transactions on 55(1):14–20

[17] Horner M, Ries L, Krapcho M, Neyman N, Aminou R, Howlader N, Altekruse S, Feuer E, Huang L, Mariotto A, et al (2009) Seer cancer statistics review, 1975–2006, National Cancer Institute. Bethesda, MD

[18] Joines WT, Zhang Y, Li C, Jirtle RL (1994a) The measured electrical properties of normal and malignant human tissues from 50 to 900 mhz. Medical physics 21(4):547–550

[19] Joines WT, Zhang Y, Li C, Jirtle RL (1994b) The measured electrical properties of normal and malignant human tissues from 50 to 900 mhz. Medical physics 21(4):547–550

[20] Jossinet J (1998) The impedivity of freshly excised human breast tissue. Physiological measurement 19(1):61

[21] Lazebnik M, McCartney L, Popovic D, Watkins CB, Lindstrom MJ, Harter J, Sewall S, Magliocco A, Booske JH, Okoniewski M, et al (2007a) A large-scale study of the ultra-wideband microwave dielectric properties of normal breast tissue obtained from reduction surgeries. Physics in medicine and biology 52(10):2637

[22] Lazebnik M, Popovic D, McCartney L, Watkins CB, Lindstrom MJ, Harter J, Sewall S, Ogilvie T, Magliocco A, Breslin TM, et al (2007b) A large-scale study of the ultrawideband microwave dielectric properties of normal, benign and malignant breast tissues obtained from cancer surgeries. Physics in Medicine and Biology 52(20):6093

[23] Malich A, Scholz B, Kott A, Facius M, Fischer D, Freesmeyer M (2007) The impact of lesion vascularisation on tumours detection by electrical impedance scanning at 200 hz. Biomedical imaging and intervention journal 3(4)

[24] Meaney PM, Fanning MW, Li D, Poplack SP, Paulsen KD (2000) A clinical prototype for active microwave imaging of the breast. Microwave Theory and Techniques, IEEE Transactions on 48(11):1841–1853

[25] Nguyen TM, Rangayyan RM (2006) Shape analysis of breast masses in mammograms via the fractal dimension. In: Engineering in Medicine and Biology Society, 2005. IEEE-EMBS 2005. 27th Annual International Conference of the, IEEE, pp 3210–3213

[26] O'Halloran M, Conceição RC, Byrne D, Glavin M, Jones E (2009) FDTD modeling of the breast: A review. Progress In Electromagnetics Research B 18:1–24

[27] Pethig R (1984) Dielectric properties of biological materials: Biophysical and medical applications. Electrical Insulation, IEEE Transactions on EI-19(5):453–474, DOI 10.1109/TEI.1984.298769

[28] Rangayyan RM, El-Faramawy NM, Desautels JL, Alim O, et al (1997) Measures of acutance and shape for classification of breast tumors. Medical Imaging, IEEE Transactions on 16(6):799–810

[29] Sha L, Ward ER, Stroy B (2002) A review of dielectric properties of normal and malignant breast tissue. In: SoutheastCon, 2002. Proceedings IEEE, IEEE, pp 457–462

[30] Stuchly M, Stuchly S (1980) Dielectric properties of biological substances–tabulated. JOURNAL OF MICROWAVE POWER

[31] Surowiec AJ, Stuchly SS, Barr JR, Swarup A (1988) Dielectric properties of breast carcinoma and the surrounding tissues. Biomedical Engineering, IEEE Transactions on 35(4):257–263

[32] Taflove A, Hagness SC (2005) Computational electrodynamics. Artech house

Chapter 3
Microwave Tomography

Tonny Rubæk and Johan Jacob Mohr

This chapter details microwave tomography for breast cancer detection. It includes a description of the scattering mechanism and introduces the object function and the forward problem, which defines the non-linear inverse tomographic problem. This is followed by a brief overview of the linear approximations which historically have been applied for solving the inverse problem. The major part of this chapter is devoted to introducing the algorithms which have been proposed for solving the non-linear tomographic problem. This includes local gradient-based methods and global evolutionary methods as well as a description of the use of multi-frequency data and a priori knowledge.

3.1 Introduction

Microwave tomography is a term used for describing methods in which measured perturbations of an electromagnetic field caused by a given object are used to derive information about the shape and electromagnetic parameters of that object.

T. Rubæk
formerly at the Department of Electrical Engineering, Technical University of Denmark, Kongens Lyngby, Denmark

OHB Systems, Bremen, Germany
e-mail: trubaek@gmail.com

J.J. Mohr (✉)
formerly at the Department of Electrical Engineering, Technical University of Denmark, Kongens Lyngby, Denmark

Mellanox Technologies, Roskilde, Denmark
e-mail: johan_jacob_mohr@yahoo.dk

© Springer International Publishing Switzerland 2016
R.C. Conceição et al. (eds.), *An Introduction to Microwave Imaging for Breast Cancer Detection*, Biological and Medical Physics, Biomedical Engineering, DOI 10.1007/978-3-319-27866-7_3

17

Tomographic methods currently used in biomedical applications include ultrasound tomography [53, 71, 72], X-ray computed tomography (CT) [1, 8, 43], positron emission tomography (PET) [4, 61], amongst others.

The use of microwave tomography dates back to the 1970s with some of the earliest reports of biomedical applications made in the beginning of the 1980s [5, 57, 59]. Compared to many of the other tomographic techniques, microwave tomography often requires more computational power since the objects under investigation are of the same size as the wavelength and the dielectric contrast are large.

In the last half of the 1990s, the spread of more affordable and powerful computers meant that the use of microwave tomography became more widespread—both for biomedical and other applications. In particular, the use of microwave tomography for breast cancer detection received massive attention in the late 1990s and early 2000s.

In this chapter, the use of microwave tomography for breast cancer detection is described. We start by introducing the integral form of the scattering problem which forms the basis for the microwave-tomography problem. After this we give a short introduction to linear microwave tomography based on the Born approximation. This type of tomography has little use in imaging for breast cancer detection but is included here for historical completeness.

Following the section on linear tomography, we describe non-linear tomography in detail in Sect. 3.4. Two different approaches are described: the gradient-based local algorithms and the evolutionary, randomised global methods. The main focus is on the more widely used gradient-based methods and the section also includes a treatment of the different regularisation schemes which have been suggested for use in microwave tomography.

Throughout this chapter, the frequency dependence is denoted by $e^{-i\omega t}$, with i being the imaginary unit and ω the angular frequency, is assumed and omitted wherever the frequency-domain phasor notation is used.

3.2 Problem Formulation

The aim of microwave tomography is to determine the distribution of the constitutive electromagnetic parameters of an object under test (OUT) positioned inside an imaging domain by means of measurements made with a number of antennas outside the imaging domain.

A schematic of a two-dimensional imaging setup is shown in Fig. 3.1. In this figure, the antennas are positioned in a circular configuration outside the imaging domain in which the OUT is completely enclosed. The field used to irradiate the domain is most commonly created by using one or more of the antennas in the measurement setup as transmitters, although other sources can also be used. By irradiating the target from multiple sources, in turn, it is possible to obtain a large number of measurements with the system.

The constitutive parameters of the object are the complex permittivity, ϵ, and the magnetic permeability, μ. In most imaging situations, it is safe to assume that the permeability of the target is equal to that of free-space, μ_0. Hence, the material

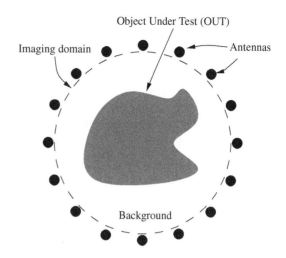

Fig. 3.1 Schematic of the tomographic imaging problem. The object under test (OUT) is positioned inside a known region in space, the imaging domain, and measurements are carried out with antennas placed outside this domain

properties at a given point $\bar{r}$ in the imaging domain are completely characterised by the complex permittivity which can be expressed using the real part of the permittivity ϵ' and the effective conductivity σ as

$$\epsilon(\bar{r},f) = \epsilon'(\bar{r},f) + i\epsilon''(\bar{r},f) = \epsilon'(\bar{r},f) + i\frac{\sigma(\bar{r},f)}{\omega}. \tag{3.1}$$

In this expression, both the permittivity and conductivity are assumed to depend on the frequency f.

The complex wavenumber is also used extensively throughout this chapter and is given as

$$k(\bar{r},f) = 2\pi f \sqrt{\epsilon(\bar{r},f)\mu_0}. \tag{3.2}$$

Above, the complex wavenumber and permittivity are given as functions of frequency. In the following sections, this frequency dependence will be ignored in order to improve readability but should be kept in mind by the reader. Also, it is worth noting that the complex permittivity of the background, i.e. everywhere outside of the OUT, is assumed to be known, and will be denoted ϵ^{bg} while the wavenumber in the background will be denoted k^{bg}.

3.2.1 Forward Problem

In order to determine the distribution of the electromagnetic parameters in the imaging domain, it is necessary to have an expression for the electromagnetic field from a given distribution. At any given position, the electric field can be expressed as

$$\bar{E}^t(\bar{r}) = \bar{E}^i(\bar{r}) + \bar{E}^s(\bar{r}) \tag{3.3}$$

wherein $\overline{E}^{\mathrm{t}}$ denotes the *total* field. The field $\overline{E}^{\mathrm{i}}$ is the incident field, i.e. the field at the observation point $\overline{r}$ when there is no scattering object present, and $\overline{E}^{\mathrm{s}}$ is the scattered field, which is defined as the additional field caused by the scattering object.

When operating in the frequency domain, the scattered field can be expressed using the object function as [10, Chap. 8]

$$\overline{E}^{\mathrm{s}}(\overline{r}) = i\omega\mu_0 \int_V \overline{\overline{G}}(\overline{r}, \overline{r}') \cdot \overline{E}^{\mathrm{t}}(\overline{r}')\chi(\overline{r}') \, \mathrm{d}v'. \tag{3.4}$$

Here, the object function χ is defined as the contrast between the scatterer and the known background

$$\chi(\overline{r}) = -i\omega \left(\epsilon(\overline{r}) - \epsilon^{\mathrm{bg}} \right). \tag{3.5}$$

The dyadic Green's function, $\overline{\overline{G}}$, is defined as the solution of the vector wave equation with a Dirac delta function as excitation, i.e.

$$\nabla \times \nabla \times \overline{\overline{G}}(\overline{r}, \overline{r}') - k^{\mathrm{bg}^2}(\overline{r})\overline{\overline{G}}(\overline{r}, \overline{r}') = \overline{\overline{I}}\delta(\overline{r} - \overline{r}') \tag{3.6}$$

where $\overline{\overline{I}}$ is the identity dyad.

The total field at any given point is thus given by the expression

$$\overline{E}^{\mathrm{t}}(\overline{r}) = \overline{E}^{\mathrm{i}}(\overline{r}) + i\omega\mu_0 \int_V \overline{\overline{G}}(\overline{r}, \overline{r}') \cdot \overline{E}^{\mathrm{t}}(\overline{r}')\chi(\overline{r}') \, \mathrm{d}v'. \tag{3.7}$$

This expression is non-linear with respect to the object function χ since the total field $\overline{E}^{\mathrm{t}}$ is present both on the left-hand side and in the integral on the right-hand side. The non-linearity of the expression is easily seen by inserting the expression for the total field in the integral on the right-hand side of the expression, resulting in

$$\overline{E}^{\mathrm{t}}(\overline{r}) = \overline{E}^{\mathrm{i}}(\overline{r}) + i\omega\mu_0 \int_V \overline{\overline{G}}(\overline{r}, \overline{r}') \cdot \overline{E}^{\mathrm{t}}(\overline{r}')\chi(\overline{r}') \, \mathrm{d}v'$$

$$= \overline{E}^{\mathrm{i}}(\overline{r}) + i\omega\mu_0 \int_V \overline{\overline{G}}(\overline{r}, \overline{r}') \tag{3.8}$$

$$\cdot \left(\overline{E}^{\mathrm{i}}(\overline{r}') + i\omega\mu_0 \int_V \overline{\overline{G}}(\overline{r}', \overline{r}'') \cdot \overline{E}^{\mathrm{t}}(\overline{r}'')\chi(\overline{r}'') \, \mathrm{d}v'' \right) \chi(\overline{r}') \, \mathrm{d}v'.$$

Given the similarity between (3.4) and the expression for the electric field radiated by a current distribution [10, Chap. 7], $\overline{J}(\overline{r})$,

$$\overline{E}(\overline{r}) = i\omega\mu_0 \int_V \overline{\overline{G}}(\overline{r}, \overline{r}') \cdot \overline{J}(\overline{r}') \, \mathrm{d}v' \tag{3.9}$$

it is straightforward to introduce the equivalent scattering current, $\overline{J}^{eq}$, as

$$\overline{J}^{eq}(\overline{r}) = \overline{E}^{t}(\overline{r})\chi(\overline{r}) \tag{3.10}$$

which leads to the expression

$$\overline{E}^{s}(\overline{r}) = i\omega\mu_0 \int_V \overline{\overline{G}}(\overline{r},\overline{r}') \cdot \overline{J}^{eq}(\overline{r}') \, dv'. \tag{3.11}$$

Hence the scattered field can be seen as the result of a radiated field from a current distribution equal to the product of the field in the imaging domain and the object function.

Expressions like (3.4) and (3.7) are known as *forward problems* and provide expressions for the measurable field as function of the distribution of the object function. However, in microwave tomography, we seek to solve the related *inverse problem*, i.e. determine the distribution of the object function from measurements of the field.

3.2.2 Measured Signals

In (3.7), the field at a given point in space is expressed as a function of the object function. However, an imaging system does not provide us with information about the field at a given point in space, but rather the signals measured on the terminals of the antennas. We can express the measured signal S^t in the same way as we did with the electric field, i.e. as a sum of an incident signal S^i, which is the signal measured when there is no scattering object present, and a scattered signal S^s, which is the change in the measured signal caused by the presence of the scattering object, i.e.

$$S^t = S^i + S^s. \tag{3.12}$$

To obtain an expression which relates the signals measured by the antennas with the distribution of dielectric parameters in the imaging domain, we apply the theorem of reciprocity, e.g. [60, Chap. 1] or [11, Chap. 1].

The theorem of reciprocity states that for two electromagnetic scenarios, scenario a and scenario b, the following relation holds true in the absence of magnetic currents

$$\int_S \left(\overline{E}_b(\overline{r}) \times \overline{H}_a(\overline{r}) - \overline{E}_a(\overline{r}) \times \overline{H}_b(\overline{r}) \right) \cdot \hat{n} \, ds$$

$$= \int_V \left(\overline{E}_a(\overline{r}) \cdot \overline{J}_b(\overline{r}) - \overline{E}_b(\overline{r}) \cdot \overline{J}_a(\overline{r}) \right) \, dv. \tag{3.13}$$

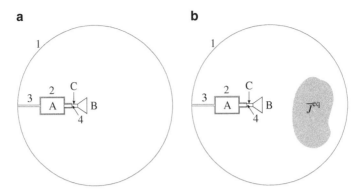

Fig. 3.2 The two different setups used for obtaining the expression for the scattered signal. In scenario *a*, the antenna is acting as a transmitter in a setup with no OUT and in scenario *b*, the antenna is acting as a receiver with the scattered field from the scatterer represented by the equivalent current $\overline{J}^{eq}$. The model of the antenna system consists of (*A*) the measurement system encased in PEC, (*B*) an antenna, and (*C*) a coaxial cable connecting the measurement system with the antenna. The integration surface is indicated by the red lines and consists of four parts: (1) a sphere with infinite radius, (2) a surface along the PEC encasement of the measurement system and outer conductor of the coaxial cable, (3) a cylinder with infinitesimal radius connecting surface 1 and 2, and (4) a surface going through the coaxial cable. (**a**) Scenario a, (**b**) Scenario b

That is, the surface integral of the difference of the cross products of the magnetic and electric fields, $\overline{E}$ and $\overline{H}$, in the two electromagnetic scenarios is equal to the volume integral of the difference of the dot products of the fields and the electric currents, $\overline{J}$, in the two scenarios. $\hat{n}$ stands for the normal vector to the surface S.

The two scenarios used for obtaining an expression for the signal as a function of the dielectric properties of the OUT are shown in Fig. 3.2. In scenario *a*, the antenna is used as a transmitter and the imaging domain is empty. In scenario *b*, the antenna is acting as a receiver while the field originates from a current distribution $\overline{J}^{eq}$, corresponding to the equivalent scattering current introduced in (3.10). The measurement hardware is assumed to be encased in an enclosure made of perfectly electrical conducting (PEC) material and the antenna connected to the measurement system via a coaxial cable. This cable is assumed to consist of a loss-less dielectric material and have inner and outer conductors made of PEC. The cable is matched to the measurement hardware but may have a mismatch with the antenna. Other configurations can be chosen, e.g. lossy materials, non-perfect conductors, and mismatch between the coaxial cable and the measurement hardware, which will make the derivations below more cumbersome but lead to the same result.

The integration surface in the two scenarios (as indicated in Fig. 3.2), indicated by the red lines in the figure, is divided into four parts.

1. The first is a sphere with infinite radius. The contribution from this part is zero since the fields in both cases satisfy the Sommerfeld radiation condition [11, Chap. 1], implying that the two cross products in the integral are equal.

Fig. 3.3 Layout of
coordinates on integration
surface 4. The antenna is
located towards $\hat{z}$

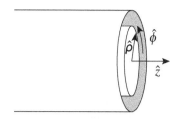

2. The second part of the surface integral covers the PEC enclosure and a part of the
   PEC outer conductor of the coaxial cable. On this surface, the electric fields have
   no tangential components, implying that the contribution from this part equals
   zero.
3. The third part of the surface integral is a thin tube which connects the integration
   surface around the PEC structure with the spherical surface with infinite radius.
   The radius of this tube may be chosen arbitrarily small, and so this part leads to
   zero contribution to the surface integral.
4. This leaves the fourth and only non-zero part of the surface integral: the plane
   covering the cross section of the coaxial cable. This part of the integration surface
   is shown in Fig. 3.3 with local cylindrical coordinates ρ, ϕ, and z.

The antenna is located towards positive z and the unit vector $\hat{z}$ is thus equal to the
negative normal vector, i.e. $\hat{z} = -\hat{n}$.

In the plane, the fields are only non-zero between the inner and outer conductors.
When the antenna is acting as a transmitter (scenario a), the fields can be expressed
as [60, Chap. 2]

$$\overline{E}^a(\rho, \phi) = V^a \frac{1 + \Gamma}{\rho \ln \frac{r^o}{r^i}} \hat{\rho} \tag{3.14a}$$

and

$$\overline{H}^a(\rho, \phi) = V^a \frac{1 - \Gamma}{2\pi \rho Z^c} \hat{\phi}. \tag{3.14b}$$

In these expressions, V^a is the voltage of the incident electric wave in the coaxial
waveguide, Γ is the reflection coefficient of the antenna defined in the plane of
integration, and Z^c is the characteristic impedance of the coaxial cable. The radius
of the inner conductor is r^i and the radius of the outer conductor is r^o.

When the antenna is operating as a receiver (in scenario b), the fields in the plane
of integration are given by

$$\overline{E}^b(\rho, \phi) = V^b \frac{1}{\rho \ln \frac{r^o}{r^i}} \hat{\rho} \tag{3.15a}$$

and

$$\overline{H}^b(\rho, \phi) = -V^b \frac{1}{2\pi\rho Z^c} \hat{\phi}$$ (3.15b)

where it has been assumed that the measurement system is matched to the coaxial line, thereby eliminating reflected waves.

The cross products of the surface integral in (3.13) can now be evaluated, leading to the following two expressions:

$$\overline{E}^a(\rho, \phi) \times \overline{H}^b(\rho, \phi) = V^a V^b \frac{1+\Gamma}{2\pi Z^c \rho^2 \ln \frac{r_o}{r_i}} \hat{n}$$ (3.16a)

and

$$\overline{E}^b(\rho, \phi) \times \overline{H}^a(\rho, \phi) = -V^a V^b \frac{1-\Gamma}{2\pi Z^c \rho^2 \ln \frac{r_o}{r_i}} \hat{n}$$ (3.16b)

which results in

$$\overline{E}^a(\rho, \phi) \times \overline{H}^b(\rho, \phi) - \overline{E}^b(\rho, \phi) \times \overline{H}^a(\rho, \phi) = V^a V^b \frac{1}{\pi Z^c \rho^2 \ln \frac{r_o}{r_i}} \hat{n}.$$ (3.16c)

The surface integral can thus be given as

$$\int_0^{2\pi} \int_{r_i}^{r_o} V^a V^b \frac{1}{\pi Z^c \rho^2 \ln \frac{r_o}{r_i}} \rho \, d\rho \, d\phi = \frac{2V^a V^b}{Z^c}.$$ (3.17)

Since there is no current in the volume in scenario a, only the first dot product in the volume integral on the right-hand side of (3.13) contributes to the integral and after rearranging the terms, we obtain

$$V^b = \frac{-Z^c}{2V^a} \int_V \overline{E}^a(\vec{r}') \cdot \overline{J}^{eq}(\vec{r}') \, dv'.$$ (3.18)

We now have an expression for the measured voltage as a function of the equivalent current distribution $\overline{J}^{eq}$. From (3.11) it is observed that the field from this distribution is equal to the scattered field and the field $\overline{E}^b$ is therefore the scattered field and V^b is the voltage measured on the receiver due to the scattered field. Using (3.10), the equivalent current can be expressed as the product of the object function χ and the field $\overline{E}^{tr}$ resulting in

$$V^b = \frac{-Z^c}{2V^a} \int_V \overline{E}^a(\vec{r}') \cdot \overline{E}^{tr}(\vec{r}') \chi(\vec{r}') \, dv'.$$ (3.19)

Here it is important to realise that the field $\overline{E}^{tr}$ is the total field composed of a known incident part and a scattered part which is equal to the field $\overline{E}^b$.

If the measurements are performed in terms of the scattered S-parameters, i.e. the ratio between the voltage of the received signal divided by the voltage V^{tr} of the excitation signal used for transmitting the field $\overline{E}^{tr}$, the expression becomes

$$S^s = \frac{-Z^c}{2V^a V^{tr}} \int_V \overline{E}^a(\overline{r}') \cdot \overline{E}^{tr}(\overline{r}') \chi(\overline{r}') \, dv'. \tag{3.20}$$

A similar expression can be found for other feeds, e.g. for a rectangular waveguide operating in TE_{10}-mode, in which case the expression is given by Dahlback et al. [13]

$$S^s_{r,t} = \frac{-Z^{TE}}{A^a A^{tr} xy} \int_V \overline{E}^a(\overline{r}') \cdot \overline{E}^{tr}(\overline{r}') \chi(\overline{r}') \, dv' \tag{3.21}$$

where Z^{TE} is the TE-wave impedance, x and y are the side lengths of the waveguide, A^a is the complex amplitude of the TE excitation used in scenario a, and A^{tr} is the complex amplitude of the TE-excitation used for generating $\overline{E}^{tr}$.

It is not possible to measure the field distribution $\overline{E}^{tr}$, since this includes the field inside of the OUT. The field $\overline{E}^a$ can be measured under certain conditions, but this is often difficult. Hence the two field distributions have to be evaluated using numerical methods.

3.3 Linear Tomography

As shown in the previous section, the expression for the scattered field and thus the contribution from the scattered field to the measured signal is non-linear with respect to the distribution of the unknown complex permittivity. This implies that obtaining the solution to the inverse problem is non-trivial.

The easiest method to overcome the challenge is by approximating the non-linear expression in (3.7) with a linear one. Different approximations have been suggested in the literature, but in this chapter we will focus on the Born approximation, which was applied in some of the earliest experiments with microwave tomography for biomedical imaging.

3.3.1 Born Approximation

One way of approximating the expression for the electromagnetic field in (3.7) is by means of a Born series [6]. The field expressed by the Born series is most easily written as

$$\overline{E}^t_M(\overline{r}) = \overline{E}^i(\overline{r}) + i\omega\mu_0 \int_V \overline{\overline{G}}(\overline{r}, \overline{r}') \cdot \overline{E}^t_{M-1}(\overline{r}') \chi(\overline{r}') \, d\overline{r}' \tag{3.22}$$

wherein $\overline{E}_M^t(\overline{r})$ is the M-order Born approximation of the field and the zero-order Born approximation is the incident field. The first-order Born approximation of the electric field is thus given by

$$\overline{E}_1^t(\overline{r}) = \overline{E}^i(\overline{r}) + i\omega\mu_0 \int_V \bar{\bar{G}}(\overline{r},\overline{r}') \cdot \overline{E}^i(\overline{r}')\chi(\overline{r}')\,\mathrm{d}\overline{r}' \tag{3.23}$$

while the second-order Born approximation is given by

$$\overline{E}_2^t(\overline{r}) = \overline{E}^i(\overline{r}) + i\omega\mu_0 \int_V \bar{\bar{G}}(\overline{r},\overline{r}') \cdot \overline{E}_1^t(\overline{r}')\chi(\overline{r}')\,\mathrm{d}v'$$

$$= \overline{E}^i(\overline{r}) + i\omega\mu_0 \int_V \bar{\bar{G}}(\overline{r},\overline{r}') \tag{3.24}$$

$$\cdot \left(\overline{E}^i(\overline{r}') + i\omega\mu_0 \int_V \bar{\bar{G}}(\overline{r}',\overline{r}'') \cdot \overline{E}^i(\overline{r}'')\chi(\overline{r}'')\,\mathrm{d}v'' \right) \chi(\overline{r}')\,\mathrm{d}v'.$$

As we can see from this expression, the use of higher-order Born approximations results in a forward problem which is non-linear with respect to the unknown object function. Hence, in order to obtain a linear expression for the total field, the first-order Born approximation must be applied.

Under the Born approximation, the expression for the scattered S-parameters given in (3.20) reduces to

$$S^s = \frac{-Z^c}{2V^aV^{tr}} \int_V \overline{E}^a(\overline{r}') \cdot \overline{E}^i(\overline{r}')\chi(\overline{r}')\,\mathrm{d}\overline{r}'. \tag{3.25}$$

where $\overline{E}^i$ is the incident field produced by the transmitting antenna. The inverse problem can thus be reduced to a Fredholm integral equation of the first kind

$$S_{r,t}^s = \int_V K_{r,t}(\overline{r}')\chi(\overline{r}')\,\mathrm{d}\overline{r}' \tag{3.26}$$

with the kernel

$$K_{r,t} = \overline{E}_r^a(\overline{r}') \cdot \overline{E}_t^i(\overline{r}'). \tag{3.27}$$

Herein, the subscripts r and t indicate different receivers and transmitters which will result in different fields in the kernel.

Numerous texts have been published on how to solve such integral equations, e.g. [3, 27, 34], but a more common, and computationally cheaper approach is to use diffraction tomography, as described in Sect. 3.3.3.

3.3.2 Validity of the Born Approximation

When we compare the expression for the first-order approximation in (3.23) with the expression in (3.7), we can see that the first-order Born approximation corresponds to approximating the total field with the incident field, i.e.

$$\overline{E}^{t}(\overline{r}) \approx \overline{E}^{i}(\overline{r}) \tag{3.28a}$$

and thus, from (3.3), the scattered field must be much weaker than the incident field

$$|\overline{E}^{s}(\overline{r})| \ll |\overline{E}^{t}(\overline{r})|. \tag{3.28b}$$

It is intuitively clear that for the scattered field to be much smaller than the incident field, the scattering object must be small and with limited contrast to the background. The limits of the contrast and size of the object in which the Born approximation is valid have been investigated by several authors [44, 70]. By analysis of the dyadic Green's function, it can be shown [10] that for the Born approximation to hold, the following relation must be satisfied for electrically small objects

$$\left| \frac{\epsilon(\overline{r})}{\epsilon^{bg}} - 1 \right| \ll 1. \tag{3.29}$$

Electrically small objects are defined as objects with a maximum extent of L for which $k^{bg}L \ll 1$. For electrically large objects, i.e. objects for which $k^{bg}L \gg 1$, the following relation must be fulfilled for the first-order Born approximation to hold [10, Chap. 8]

$$k^{bg}L \left| \frac{\epsilon(\overline{r})}{\epsilon^{bg}} - 1 \right| \ll 1. \tag{3.30}$$

From this expression (3.30) and (3.29), it is observed that the requirements for the contrast of electrically large objects are much stricter than that for the contrast of electrically small objects.

3.3.3 Diffraction Tomography

Although the inverse microwave tomography problem based on the first-order Born approximation can be approached as a Fredholm integral equation of the first kind and solved using one of the numerous standard methods, a more widespread approach for solving the problem is the so-called diffraction tomography.

With this method, $\overline{E}_r^a$ and $\overline{E}_t^i$ are represented using their wave spectra, i.e. the Fourier transform of the fields under the assumption that the OUT is in the far field of the antennas. The resulting expressions map the measured data to a Fourier-domain representation of the OUT, i.e. the spatial Fourier transform of the object function $\chi(\overline{r})$.

These wave spectra can, for example, be expressed using the plane-wave spectrum, the cylindrical-wave spectrum, or the spherical-wave spectrum [2, 35, 62], depending on what is best suited for the topology used in the imaging system. In an ideal illumination and measurement setup both transmitters and receivers completely surround the imaging domain. In that case, the measurements map to a volume in the three-dimensional Fourier domain corresponding to a sphere with radius $2k^{bg}$ centred at the origin. A low-pass filtered version of the distribution of χ can then be found by performing an inverse Fourier transform on the data. For non-perfect illuminations and measurements, the volume in the Fourier domain is not completely filled, but by imposing restrictions on $\chi(\overline{r})$ and/or by using multiple frequencies this can be mitigated to some extent.

Diffraction tomography has been suggested for breast cancer detection [30, 31] and other biomedical applications [42]—including some of the first published biomedical applications [5, 57, 59]. In general, these methods suffer from the fact that the objects must satisfy the requirements set in (3.29) and (3.30), i.e. the scattering object must be small and have low contrast. This is rarely the case in biomedical applications of diffraction tomography, but by imposing a priori knowledge, it is still possible to obtain useful results for some simple problems.

3.4   Non-Linear Microwave Tomography

Although the approaches described in the previous section have been successfully applied for creating tomographic images in low-contrast scenarios, they are of limited use for biomedical imaging where the OUT is often highly heterogeneous with large regions of unknown permittivity. In such cases, the simplification of assuming that the total field can be approximated as the known incident field is not accurate and the inverse tomography problem cannot be linearised. Instead, the problem must be solved as a non-linear problem. Solving this non-linear problem is difficult because the problem is ill-posed in the classical sense. That is, the inverse problem has no unique solution and the solution is not stable, i.e. does not depend continuously on the measured data. Furthermore, the presence of noise and modelling errors implies that a solution might not exist.

To solve the problem, we first discretise the imaging domain into Q subvolumes in each of which the object function is assumed to be constant, i.e.

$$\chi(\overline{r}) = \chi_q \quad \text{for} \quad \overline{r} \in V_q \tag{3.31}$$

where V_q ($q \in [1; Q]$) is the qth subvolume. The distribution of the object function in the imaging domain can thus be written as a Q-element vector $\overline{\chi}$ in which each element corresponds to a subvolume in the imaging domain.

Two different approaches can be used for solving such non-linear inversion problems: gradient-based local algorithms and global algorithms. In both types of algorithms, the tomographic problem is formulated as a minimisation problem in which a cost function is sought to be minimised through an iterative process. The imaging problem thus reduces to

$$\overline{\chi} = \arg \min \{|F(\overline{\chi})|\} \tag{3.32}$$

where F is the cost function and the vector $\overline{\chi}$ is a vector holding the contrast in each of the subvolumes in the imaging domain. The cost function can be expressed as

$$F(\overline{\chi}) = \left(F^{\mathrm{d}}(\overline{\chi}) + F^{\mathrm{ra}}(\overline{\chi})\right) F^{\mathrm{rm}}(\overline{\chi}) \tag{3.33}$$

where $F^{\mathrm{d}}(\overline{\chi})$ is a term related to the measured data, $F^{\mathrm{ra}}(\overline{\chi})$ is an additive regularisation term, and $F^{\mathrm{rm}}(\overline{\chi})$ is a multiplicative regularisation term.

The term related to the measured data, F^{d}, is given as the difference between the measured data and the corresponding calculated data, i.e.

$$F^{\mathrm{d}}(\overline{\chi}) = \sum_{t=1}^{T} \sum_{r=1}^{R} |S_{r,t}^{\mathrm{m}} - S_{r,t}^{\mathrm{c}}(\overline{\chi})|^2 \tag{3.34}$$

where $S_{r,t}^{\mathrm{m}}$ is the measured signal for the antenna combination with receiver r and transmitter t with total number of transmitters T and receivers R. The corresponding calculated signal, when the distribution of the object function $\overline{\chi}$ is present in the system, is given by $S_{r,t}^{\mathrm{c}}$.

The cost function is often expressed in vector form as

$$F(\overline{\chi}) = \left\| \begin{bmatrix} \overline{F}^{\mathrm{d}}(\overline{\chi}) \\ \overline{F}^{\mathrm{ra}}(\overline{\chi}) \end{bmatrix} \right\|_2^2 \left\| \overline{F}^{\mathrm{rm}}(\overline{\chi}) \right\|_2^2 = \|\overline{F}\|_2^2. \tag{3.35a}$$

Here, $\overline{F}^{\mathrm{d}}$ is the data vector which holds the difference between the measured and calculated signals, i.e.

$$\overline{F}^{\mathrm{d}}(\overline{\chi}) = \overline{S}^{\mathrm{m}} - \overline{S}^{\mathrm{c}}(\overline{\chi}). \tag{3.36}$$

The minimisation problem can be solved by either using local, gradient-based methods or global methods. With local methods, a solution is sought in the space around an initial distribution $\overline{\chi}_0$, by iteratively linearising the cost function on the basis of the gradient of the function. With global methods, we do not constrain ourselves by using the gradient, but attempt instead to search the entire solution space, e.g. by using particle swarm or genetic algorithms. However, for each new proposed solution in these methods, it is necessary to calculate the elements in the vector $\overline{S}^c$. For three-dimensional tomography systems this can be a time-consuming task, lasting hours, even on high-performance computers. Therefore, the most widespread methods are the gradient-based—even though these have the drawback of being susceptible to getting trapped in local, erroneous minima which the global methods avoid. Hence, the majority of this section will be used to describe gradient methods, while a brief review of global methods is given in Sect. 3.4.5.

The initial distribution $\overline{\chi}_0$ is most often chosen to be the empty system, that is, the object function is set to zero throughout the imaging domain. However, schemes have been proposed in which a priori knowledge about the object function is used to obtain an initial distribution closer to the assumed solution.

In the local gradient-based methods, which are also known as Newton methods, the cost function is minimised through an iterative process. In the most basic form, the object function is updated in each iteration $p \in [1, 2, 3, \ldots, P]$ using

$$\overline{\chi}_{p+1} = \overline{\chi}_p + \alpha \overline{\Delta \chi}_p \tag{3.37}$$

where α is the scalar Newton step with a value between 0 and 1. Under the assumption that the cost function can be approximated as

$$\overline{F}_{p+1} \approx \overline{F}_p + \underline{\underline{J}}_p \overline{\Delta \chi}_p \tag{3.38}$$

where $\underline{\underline{J}}$ is the Jacobian matrix holding the first-order partial derivatives of the cost function with respect to the elements in $\overline{\chi}_p$. The update vector $\overline{\Delta \chi}_p$ can thus be found as

$$\overline{\Delta \chi}_p = -\underline{\underline{J}}_p^{-1} \overline{F}_p \tag{3.39}$$

where $\underline{\underline{J}}_p^{-1}$ is the inverse of the Jacobian matrix.

The use of this simple algorithm for determining the update is, of course, only possible when the Jacobian matrix can be inverted, which is not always the case. And in cases where the matrix is invertible, it might not be the optimum way to determine the update. Different methods for determining the update are described below.

3.4.1 Determining the Update: Derivative of the Measured Signal

In order to linearise the cost function, it is necessary to find the derivative of the measured scattered signal with respect to a change in the object function in a single subvolume. By following a procedure similar to that used in Sect. 3.2.2 to determine an expression for the scattered signal, the partial derivative of the scattered signal with respect to the object function in a given subvolume of the imaging domain is found to be

$$\frac{\partial S_{r,t}}{\partial \chi_q} = \frac{\partial S^s_{r,t}}{\partial \chi_q} = \frac{-Z^c}{2V^{rec}V^{tr}} \int_{V_q} \overline{E}^{rec}(\vec{r}') \cdot \overline{E}^{tr}(\vec{r}') \, dv'. \qquad (3.40)$$

In this expression, $\overline{E}^{rec}$ is the field in the subvolume V_q when the receiver is transmitting with an excitation voltage of V^{rec}, and $\overline{E}^{tr}$ is the field in the imaging domain when the transmitter is transmitting with excitation voltage V^{tr}. Both of these fields are the fields in the imaging domain when the distribution of the object function $\overline{\chi}$ is present in the domain, and not the empty system which is used in the expression for calculating the scattered field in (3.20).

3.4.2 Regularisation of the Calculation of Updates in Gradient Methods

The most well-known implementation of a gradient-based method is the Gauss–Newton method [54, Chap. 8] in which the distribution of the object function is updated using

$$\overline{\chi}_{p+1} = \overline{\chi}_p - \left(\underline{\underline{J}}_p^H \underline{\underline{J}}_p \right)^{-1} \underline{\underline{J}}_p^H \overline{F}_p, \qquad (3.41)$$

wherein $\underline{\underline{J}}_p^H$ is the conjugate transpose of the Jacobian matrix. This implies that the algorithm determines the update by solving the normal equations of the problem in (3.39). Since the matrix $\underline{\underline{J}}_p^H \underline{\underline{J}}_p$ is symmetric and positive semi-definite, this expression can be used in many cases where the inverse of $\underline{\underline{J}}_p$ is non-existing and it is therefore impossible to use the simple expression in (3.39).

In microwave tomography algorithms it is often the case that the explicit regularisation terms F^{ra} and F^{rm} are omitted from the cost function (3.33), leaving only the data-related F^d in the cost function. The regularisation of the inverse problem is then obtained instead by applying regularisation in the calculation of the update $\overline{\Delta \chi}$, i.e. in the expressions in (3.39) and (3.41). Such methods are sometimes referred to as *inexact* or *modified* Newton algorithms [14, 19, 20, 54].

3.4.2.1 Tikhonov Regularisation

One of the most commonly applied regularisation schemes in linear inverse problems is the Tikhonov regularisation [58, 73] and as a result of its widespread use, it has also been applied for regularising the update problem in non-linear microwave tomography by several authors.

The general form of the Tikhonov regularisation leads to the following minimisation problem which must be solved in order to obtain the update of the object function

$$\overline{\Delta \chi}_p = \arg\min \left\{ \left\| \underline{\underline{J}}_p \overline{\Delta \chi}_p + \overline{F}_p \right\|_2^2 + \lambda^2 \left\| \underline{\underline{L}} \overline{\Delta \chi}_p \right\|_2^2 \right\}$$

$$= \arg\min \left\{ \left\| \begin{bmatrix} \underline{\underline{J}}_p \\ \lambda \underline{\underline{L}}_p \end{bmatrix} \overline{\Delta \chi}_p + \begin{bmatrix} \overline{F}_p \\ \overline{0} \end{bmatrix} \right\|_2^2 \right\} \qquad (3.42)$$

where the regularisation parameter λ is a positive scalar, $\underline{\underline{L}}$ is the regularisation matrix, and $\overline{0}$ is a zero vector. The update to the non-linear problem is then found as

$$\overline{\Delta \chi}_p = \left(\underline{\underline{J}}_p^H \underline{\underline{J}}_p + \lambda^2 \underline{\underline{L}}^H \underline{\underline{L}} \right)^{-1} \underline{\underline{J}}_p^H \overline{F}_p. \qquad (3.43)$$

Most often the identity matrix $\underline{\underline{I}}$ is used as the regularisation matrix so that $\underline{\underline{L}}^H \underline{\underline{L}} = \underline{\underline{I}}$, in which case the Tikhonov algorithm simply penalises the norm of the update vector.

When Tikhonov regularisation is applied in the calculation of the update vector, it is important to choose the right value of the regularisation parameter λ. If the value of λ is too high, the problem is over-regularised and the solution will be smooth and lack detail. On the other hand, if the value of λ is too small, regularisation will be without effect.

A number of different approaches have been described in the literature for determining λ when dealing with linear problems, e.g. the L-curve criterion and generalised cross validation [34, Chap. 7]. However, in most of the published literature on non-linear microwave tomography, the value of the regularisation parameter has been chosen *ad hoc*, i.e. by testing the inversion algorithm with different values of λ and choosing the one which gives the best result.

This is, in part, a result of the fact that the optimum solution to the linear problem might not be the optimum solution when solving for the update for a non-linear problem. Here, over-regularised solutions in the first iterations might lead to better results than if the "optimal" solutions with respect to the linear problem are used.

Slightly modified versions of the Tikhonov regularisation for solving non linear problems are the Levenberg [47] and Levenberg–Marquardt algorithms [48]. In the Levenberg algorithm, the regularisation matrix is the identity matrix and the regularisation parameter is decreased as the algorithm progresses, which leads to the following linear equation for determining the update vector:

$$\overline{\Delta \chi}_p = \arg\min \left\{ \left\| \begin{bmatrix} \underline{\underline{J}}_p \\ \lambda_p \underline{\underline{I}} \end{bmatrix} \overline{\Delta \chi}_p + \begin{bmatrix} \overline{F}_p \\ \overline{0} \end{bmatrix} \right\|_2^2 \right\} \qquad (3.44)$$

and thus

$$\overline{\Delta \chi}_p = \left(\underline{\underline{J}}_p^H \underline{\underline{J}}_p + \lambda_p^2 \underline{\underline{I}}\right)^{-1} \underline{\underline{J}}_p^H \overline{F}_p. \tag{3.45}$$

When using this approach, the value of λ_p is set to a high value in the first few iterations of the imaging algorithm and then decreased as the algorithm progresses, i.e. $\lambda_p > \lambda_{p+1} > \lambda_{p+2} > \cdots > \lambda_P$. The reasoning behind this is that as the imaging algorithm progresses and the reconstructed distribution $\overline{\chi}_p$ gets closer to the actual distribution in the imaging domain, the linearisation of the problem becomes a better approximation of the problem, which in turn means that less regularisation is needed to obtain good results for the update.

The Levenberg–Marquardt algorithm further refines this procedure by using the regularisation matrix

$$\underline{\underline{L}}_p = \mathrm{diag}(\underline{\underline{J}}^H \underline{\underline{J}}) \tag{3.46}$$

instead of the identity matrix $\underline{\underline{I}}$ used in (3.44), and $\underline{\underline{L}}_p^H \underline{\underline{L}}$ instead of $\underline{\underline{I}}$ in (3.45). This algorithm has successfully been applied for biomedical microwave tomography, including imaging of the breast [50].

As is the case with the Tikhonov regularisation scheme, the choice of λ_p is critical to the performance of the Levenberg and Levenberg–Marquardt algorithms. In these, the regularisation parameter λ_p is determined on the basis of the error in successive iterations and decreased as the algorithm progresses. However, there has not been any stable general-purpose method proposed to determining the optimal values of the parameters. Instead, *ad hoc* methods were applied.

3.4.2.2 Iterative Algorithms and Spectral Methods

Instead of explicitly choosing the regularisation parameter in the Tikhonov algorithm, the use of iterative algorithms such as the conjugate gradient least squares (CGLS) algorithm [36], the least square QR (LSQR) factorisation algorithm [55], the Landweber algorithm [46], and the algebraic reconstruction technique (ART) algorithm [29] have also been applied to microwave tomography. Their advantage is that it is not necessary to solve the large linear problems resulting from the Tikhonov-based regularisation schemes, i.e. calculating the inverse of the matrices involved in (3.43) and (3.45). Instead, the iterative algorithms solve the problem through a number of computationally inexpensive matrix-vector multiplications and the regularising effect of the algorithms is controlled by varying the number of iterations the algorithms are allowed to run with fewer iterations corresponding to more regularisation. When discussing these iterative algorithms, it is important to distinguish between the iterations of the gradient-based algorithms, which is denoted $p \in [1, P]$ in this text, and the iterations of the algorithms used to calculate the update, which will be denoted $m \in [1, M]$, in each of the P iterations.

As an example, the iterative Landweber algorithm finds the solution to the linear update problem as

$$\overline{\Delta\chi}_p^{[m]} = \overline{\Delta\chi}_p^{[m-1]} + \beta\underline{\underline{J}}_p^H\left(\overline{F}_p - \underline{\underline{J}}_p\overline{\Delta\chi}_p^{[m-1]}\right) \tag{3.47}$$

where m is the Landweber iteration number, β is a positive real number satisfying

$$0 < \beta < \frac{2}{\left\|\underline{\underline{J}}^H\underline{\underline{J}}\right\|_2}, \tag{3.48}$$

and $\overline{\Delta\chi}_p^{[0]}$ is equal to the zero vector $\overline{0}$.

Another example of an iterative algorithm is the CGLS algorithm. In this, the solution is updated in each of the M iterations using the following five equations:

$$\gamma = \frac{\left\|\underline{\underline{J}}_p^H\overline{a}^{[m-1]}\right\|_2^2}{\left\|\underline{\underline{J}}_p\overline{d}^{[m-1]}\right\|_2^2} \tag{3.49a}$$

$$\overline{\Delta\chi}_p^{[m]} = \overline{\Delta\chi}_p^{[m-1]} + \gamma^{[m]}\overline{d}^{[m-1]} \tag{3.49b}$$

$$\overline{a}^{[m]} = \overline{a}^{[m-1]} - \gamma^{[m]}\underline{\underline{J}}_p\overline{d}^{[m-1]} \tag{3.49c}$$

$$\beta^{[m]} = \frac{\left\|\underline{\underline{J}}_p^H\overline{a}^{[m]}\right\|_2^2}{\left\|\underline{\underline{J}}_p^H\overline{a}^{[m-1]}\right\|_2^2} \tag{3.49d}$$

$$\overline{d}^{[m]} = \underline{\underline{J}}_p^H\overline{a}^{[m]} + \beta^{[m]}\overline{d}^{[m-1]} \tag{3.49e}$$

with the vectors $\overline{\Delta\chi}_p, \overline{a}$, and $\overline{d}$ initialised as

$$\overline{\Delta\chi}_p^{[0]} = \overline{0} \tag{3.50a}$$

$$\overline{a}^{[0]} = \overline{F}_p - \underline{\underline{J}}_p\overline{\Delta\chi}_p^{[0]} \tag{3.50b}$$

$$\overline{d}^{[0]} = \underline{\underline{J}}_p^H\overline{a}^{[0]}. \tag{3.50c}$$

The solution obtained with the CGLS algorithm is the solution to the normal equation

$$\overline{\Delta\chi}_p = \arg\min\left\|\underline{\underline{J}}_p^H\underline{\underline{J}}_p\overline{\Delta\chi}_p + \overline{F}_p\right\|_2 \tag{3.51}$$

subject to $\overline{\Delta\chi}$ being spanned by the first m Krylow vectors of the problem. In terms of computational complexity and scope, the LSQR algorithm is very similar to the CGLS algorithm with the difference that the solution in the LSQR algorithm is spanned by the first m Lanczos vectors of the problem.

Similar to the Tikhonov regularisation schemes, in which the performance is governed by the choice of regularisation parameter λ, the performance of the iterative schemes are governed by the number of iterations M. Different approaches have been proposed for choosing the number of iterations, including the use of the L-curve [69] and a scheme in which the number M of iterations is increased as the algorithm progresses, leading to less regularisation in the latter Newton iterations compared to the first [63].

When we try to determine the optimum regularisation, the iterative methods have the advantage, over the Tikhonov-based algorithms, that the different levels of regularisation are provided one by one as m increases. This implies that by simply storing the update vectors $\overline{\Delta\chi_p}^{[m]}$, M updates of different levels of regularisation are obtained and are available for analysis. For the Tikhonov regularisation schemes, a full matrix inversion must be performed for each different value of λ, making it more computationally expensive to investigate different levels of regularisation.

Also, *ad hoc* approaches in which the number of iterations is determined by imperative trial-and-error have been applied successfully for tomographic imaging.

Closely related to the CGLS and LSQR algorithms are the spectral methods in which the solution is regularised by projection onto a subspace spanned by a pre-selected set of vectors instead of the subspaces spanned by the Krylow or Lanczos vectors. Different sets have been proposed for tomographic imaging of the breast, including Fourier series [67, 68], Gaussian-based subspaces [74], and subspaces spanned by wavelets [65]. These subspaces can be chosen in such a way that information about the structure of the breast is incorporated into the vectors or the subspace may be chosen in such a way that the vectors do not contain spatial details smaller than what is the expected resolution limit of the imaging system. In either case, the regularising effect originates from the choice of the subspace which, in turn, implies that it is important to choose a proper set of vectors to span this subspace.

3.4.3 Regularisation Terms in the Cost Function

Previously, we only considered the regularisation of the linear problem related to finding the update in the gradient-based methods. However, as indicated in (3.33), regularising terms can also be implemented in the cost function itself. Different additive and multiplicative terms have been suggested. Both additive and multiplicative terms are based on assumptions made on the distribution of the (unknown) object function. Some of the proposed methods are described below.

In [49], the authors use a two-stage approach in which an additive term is introduced halfway through the inversion. In the first iterations of the Newton algorithm, the cost function

$$F(\overline{\chi}_p) = F^{\mathrm{d}}(\overline{\chi}_p) \tag{3.52}$$

is used, but once an intermediate estimate of the object function $\overline{\chi}_{\mathrm{IM}}$ is found, a Euclidean-distance penalty term is introduced. This leads to the new cost function

$$F^{\mathrm{Euclid}}(\overline{\chi}_p) = F^{\mathrm{d}}(\overline{\chi}_p) + F^{\mathrm{ra}}(\overline{\chi}_p, \overline{\chi}_{\mathrm{IM}}) \tag{3.53a}$$

with

$$F^{\mathrm{ra}} = \rho \left\| \overline{\chi}_p - \overline{\chi}_{\mathrm{IM}} \right\|_2^2 \tag{3.53b}$$

where ρ is a positive weighting parameter. The idea behind this approach is that the introduction of the Euclidean-distance penalty term allows for less regularisation in the Tikhonov algorithm used to determine the update vector $\overline{\Delta \chi}_p$. In turn, this allows for reconstruction of finer details in the images.

Another approach is taken in [51], where the authors apply the multiplicative term[1]

$$F^{\mathrm{rm}}(\overline{\chi}_p) = \frac{1}{V} \int\limits_V \frac{|\nabla \chi_p(\vec{r})|^2 + \delta_p^2}{|\nabla \chi_{p-1}(\vec{r})|^2 + \delta_p^2} \, \mathrm{d}v' \tag{3.54}$$

to the cost function with δ_n being a positive parameter. This term allows the algorithm to better preserve the edges of objects in the final image by countering the smoothing effects introduced by typical least-squares solvers mentioned above. This implies that (3.54) can produce superior reconstructions when the distribution of the object function has sharp edges. This approach is closely related to another algorithm proposed by the same authors [52], in which they assume that the object function can only take on a finite number L of discrete values χ_l and obtain the multiplicative regularisation term[2]

$$F^{\mathrm{rm}}(\overline{\chi}_p) = \frac{1}{V} \int\limits_V \prod_{l=1}^{L} \frac{|\chi_p(\vec{r}) - \chi_l|^2 + \delta_p^2}{|\chi_{p-1}(\vec{r}) - \chi_l|^2 + \delta_p^2} \, \mathrm{d}v'. \tag{3.55}$$

---

[1]In [51], the authors deal with 2-D tomography. The expression presented here has been re-written to fit a 3-D algorithm.

[2]In [52], the authors deal with 2-D tomography. The expression presented here has been re-written to fit a 3-D algorithm.

This term also leads to reconstructions in which the edges of the individual objects are more well defined, but requires that the possible values of the contrast function is known a priori. Other algorithms based on level-set methods have also been proposed [12, 17, 33, 40, 41] and have all yielded good results for those cases where the values of the contrast function are known in advance.

Although some algorithms have been proposed in which additive and multiplicative regularising terms are used in the cost function, the majority of published algorithms do not use these terms. And when these terms are included, additional regularisation is applied when the updates are determined, as described above. Hence, the regularising terms in the cost function should not be viewed as a substitute for the regularisation applied during the calculation of the update, but rather as a supplement.

3.4.4 Multiple Frequencies and Time-Domain Tomography

Above, only single-frequency algorithms were considered. However, the non-linear tomography algorithms are easily extended to using data from multiple frequencies. One approach is called the stepped-frequency method, but is also known as the march-on-frequency or frequency-hopping method. Another approach is the multi-frequency method. Here, a few examples of both methods will be described.

In the stepped-frequency method, measurement data from multiple frequencies are used by successively solving the tomography problem at different frequencies starting with the lowest. The solution obtained at one frequency, $\overline{\chi}_{P,f}$ (with $f \in [1, \ldots, F]$ being the frequency), is then used as the starting point for the reconstruction at the next frequency, $\overline{\chi}_{0,f_{n+1}}$. The reasoning is that at the lower frequencies, the tomographic algorithm is more likely to reach a solution which is close to the actual distribution of the object function in the imaging domain. However, this comes at the cost of a loss of detail in the distribution compared to the higher frequencies.

The distribution obtained at the lower frequency should be closer to the actual distribution than the empty system. Hence, by using the distribution as the starting point for the algorithm at the higher frequency, a gradient-based algorithm, which only searches for a solution in a limited space around the initial distribution, should have a better chance of finding the correct distribution.

Alternatively, data from multiple frequencies can be included in a multi-frequency approach in which all the data is included in the minimisation problem at the same time. When this is done, the data vector $\overline{F}^{d}$ is augmented with data from multiple frequencies, leading to

$$\overline{F}^{\mathrm{d}} = \begin{bmatrix} \overline{F}^{\mathrm{d}}_1 \\ \vdots \\ \overline{F}^{\mathrm{d}}_f \\ \vdots \\ \overline{F}^{\mathrm{d}}_F \end{bmatrix}. \tag{3.56}$$

The idea behind this approach is that the inclusion of data from multiple frequencies should make the gradient-based algorithms more likely to search in the correct direction. Also, the algorithm should be less likely to end up in local erroneous minima, as the solution which is a minimum for one frequency is not likely to be the solution for other frequencies used in the inversion.

A key issue when using multiple frequencies is how to handle the frequency dependency of the object function $\overline{\chi}$, i.e. the frequency dependence of the complex permittivity. This is particularly important when using microwave imaging for biomedical applications, since the permittivity of most biological tissues is frequency dependent. A popular choice, able to model the frequency behaviour of most relevant tissue types, is the single-pole Debye model. It expresses the complex permittivity as a function of the four real-valued parameters: permittivity at ∞ frequency (ϵ^∞), $\Delta\epsilon$ is the variation between static permittivity (ϵ^{s}) and ϵ^∞, time of relaxation (τ), and static conductivity (σ^{s}) as

$$\epsilon(f) = \epsilon^\infty + \frac{\Delta\epsilon}{1 - i2\pi f \tau} - \frac{\sigma^{\mathrm{s}}}{i2\pi f}. \tag{3.57}$$

The tomographic microwave algorithm can then be reformulated from using the object function χ to using the parameters in the chosen model by means of the chain rule and (3.40), e.g.

$$\frac{\partial S_{r,t}}{\partial \Delta\epsilon_q} = \frac{\partial S_{r,t}}{\partial \chi_q} \frac{\partial \chi_q}{\partial \Delta\epsilon_q} = \frac{1}{1 - i2\pi f \tau_q} \frac{-Z^{\mathrm{c}}}{2V^{\mathrm{rec}}V^{\mathrm{tr}}} \int_{V_q} \overline{E}^{\mathrm{rec}}(\overline{r}') \cdot \overline{E}^{\mathrm{tr}}(\overline{r}') \, \mathrm{d}v' \tag{3.58}$$

where subscript q in $\Delta\epsilon$ and τ indicates that these are the parameters for sub-volume V_q. This implies that when the Debye model is used for modelling the frequency dependence, each subvolume in the imaging domain has four real-valued parameters that have to be reconstructed instead of the single complex parameter χ_q.

An additional advantage of the single-pole Debye model is that it is easy to implement in the finite difference time-domain (FDTD) method, which is a popular computational method for solving the forward problem when using multiple frequencies in the tomographic algorithms. Models similar to the single-pole Debye model, such as the Cole–Cole model, have also been proposed for tomographic imaging.

Simpler models than the Cole–Cole and single-pole Debye models have also been proposed for multi-frequency imaging. In [23], the authors model the frequency dependency of the real and imaginary parts of the complex permittivity as a simple linear function of the frequency. This results in four parameters to be reconstructed for each subvolume, as is the case with the single-pole Debye model.

An even simpler way of dealing with the frequency dependency is to simply ignore it and assume that the complex permittivity is constant over frequency. This approach has been successfully applied for stepped-frequency algorithms wherein the change in permittivity from one frequency to the next is limited. This implies that the reconstructed distribution at a lower frequency may still be a better starting point for the reconstruction than the empty system, as long as the permittivity changes only slightly with frequency.

Closely related to multi-frequency tomography is time-domain tomography. Here, all signals in the algorithm are time-domain signals rather than frequency-domain signals as used above. Most commonly, the cost function used when doing time-domain tomography is expressed as

$$F(\overline{\chi}) = \sum_{t=1}^{T} \sum_{r=1}^{R} \int_{0}^{\Psi} |S_{r,t}^{m}(\tau) - S_{r,t}^{c}(\tau, \overline{\chi})|^2 \, d\tau = \sum_{t=1}^{T} \sum_{r=1}^{R} F_{r,t}. \tag{3.59}$$

In this expression, both S^m and S^c are time-domain signals and are a function of the time variable τ which runs from 0 to the total measurement time Ψ. The partial derivatives of the cost function with respect to the real and imaginary parts of the object function are given by Gustafsson and He [32]

$$\frac{\partial F_{r,t}}{\partial \operatorname{Re}(\chi_q)} = -2\eta_{bg} \int_{V_q} \int_{0}^{\Psi} \overline{D}_{r,t}(\tau, \overline{r}') \cdot \overline{E}_t^{t}(\tau, \overline{r}') \, d\tau \, dv' \tag{3.60a}$$

and

$$\frac{\partial F_{r,t}}{\partial \operatorname{Im} \chi_q} = \frac{2\eta_{bg}}{i\omega} \int_{V_q} \int_{0}^{\Psi} \overline{D}_{r,t}(\tau, \overline{r}') \cdot \partial_\tau \overline{E}_t^{tr}(\tau, \overline{r}') \, d\tau \, dv' \tag{3.60b}$$

where η_{bg} denotes wave impedance of the background. In these expressions, $\overline{E}_t^{tr}(\tau, \overline{r}')$ is the field in the imaging domain produced by the transmitting antenna, $\overline{D}_{r,t}$ is the field produced by the receiving antenna when it is excited with the time-reversed difference between the measured and simulated signals. The excitation signal $S_{r,t}^{exct}$ is given by

$$S_{r,t}^{exct}(\tau) = S_{r,t}^{m}(\Psi - \tau) - S_{r,t}^{c}(\Psi - \tau). \tag{3.61}$$

It should also be noted that the partial derivative in (3.60b) involves the partial derivative of the field. The equations in (3.60) allow for formulating a Jacobian matrix for the minimisation of the cost function and the solution of the problem can thus be solved using the same techniques as applied in the frequency-domain algorithms.

Time-domain tomography is applied less often than frequency-domain tomography but it has been successfully applied for reconstructing images of both numerical and experimental phantoms, e.g. in [24, 25, 75]

A key issue in both time-domain and multi-frequency tomography is the weighting of the data obtained at different frequencies. In time-domain tomography the weighting of the different frequency components is done implicitly by choosing the shape of the time-domain pulse used to excite the transmitting antenna [24] while the weight of the individual frequency components in a multi-frequency algorithm can be chosen explicitly by simply multiplying each data from each frequency in (3.56).

Since the high-frequency components are attenuated more than the low-frequency components of the field when they propagate through the breast, it can be useful to give the high-frequency data more weight in the reconstruction. Otherwise, the reconstruction algorithm may ignore these data simply because their contribution to the total value of the cost function is smaller than the contribution from the low-frequency components.

3.4.5 Global Methods

In addition to the gradient-based methods, global methods have also been suggested for solving the tomographic problem. The suggested methods have primarily been based on evolutionary algorithms such as genetic algorithms [37, 38, 66] or different types of particle-swarm optimisation (PSO) algorithms [7, 18, 45].

The major advantage of the evolutionary algorithms is that they are less likely to get trapped in a local minimum than the gradient-based algorithms, since the former search the entire solution domain and are not limited to search only in the direction of the gradient. However, this comes at the cost of being much more computationally expensive than the gradient-based algorithms—often to the extent where the computational demands make them prohibitively slow.

Different approaches for simplifying the problem and thereby reducing the computational complexity of the problem have been proposed. The most widespread are to do only two-dimensional imaging and to reduce the number of cells in the imaging domain. These approaches have been successfully used for recreating images from synthetic data by several authors, e.g. [9, 26, 64].

Some attempts to solve the three-dimensional tomography problem using evolutionary algorithms have also been presented. In all of these, the authors have introduced some sort of constraints on the problem in order to reduce the complexity.

In [16], the authors simplify the problem by assuming that the shape and dielectric properties of the breast is known and that the only unknown quantities which need to be reconstructed are the position of the center, the radius, and the complex permittivity of a single spherical tumour. A different approach to simplify the problem was introduced in [15]. Herein, the authors use the so-called multi-scaling technique to reduce the dimensionality of the inverse problem which allows for the use of a PSO algorithm. Similarly, the authors in [39] propose a method for reducing the population size in a PSO algorithm, thereby reducing the overall number of forward problems which needs to be solved.

3.4.6   Use of a Priori Knowledge

As the last point in this section, we discuss the use of a priori knowledge in combination with the non-linear tomographic algorithms. One of the most common ways of applying a priori knowledge is to introduce an initial distribution $\overline{\chi}_0$ for the algorithm, which is different from the uniform background and/or limits the size of the imaging domain so that it only covers the breast. By choosing a starting point which is closer to the actual distribution, the chance of the algorithm getting stuck in erroneous local minima is reduced.

The initial guess can be obtain in different ways, e.g. by laser [56] measurements of the breast or by means of a magnetic resonance imaging (MRI) scan [22]. Once the position of the surface of the breast has been obtained it is straightforward to limit the imaging domain to cover only the volume occupied by the breast. It is also straightforward to set the initial constitutive parameters of this region to expected values of the breast tissue (instead of using the constitutive parameters of the background) and to even include a skin layer at a fixed position in the model. If these initial values are chosen properly, this will all bring the initial distribution closer to the actual distribution and, in turn, reduce the risk of getting trapped in erroneous minima.

Other forms of a priori knowledge which have been suggested includes the use of the level-set algorithms mentioned above and the regularisation term in (3.55) which requires a priori knowledge of the possible values of the object function. Also, the use of structural information about the inside of the breast, e.g. obtained from MRI, has been suggested [21, 28]. Although the initial distribution in these algorithms may still be the uniform background, the fact that the solution space is limited, either by imposing restrictions on the value χ or by limiting the degrees of structural freedom, implies that the reconstruction has a better chance of ending in a suitable minimum.

References

[1] Alvarez RE, Macovski A (1976) Energy-selective reconstructions in X-ray computerised tomography. Phys Med Biol 21(5):733, DOI 10.1088/0031-9155/21/5/002

[2] Anastasio MA, Shi D, Huang Y, Gbur G (2005) Image reconstruction in spherical-wave intensity diffraction tomography. J Opt Soc Am A 22(12):2651–2661, DOI 10.1364/JOSAA.22.002651

[3] Atkinson KE (1976) A survey of numerical methods for the solution of Fredholm integral equations of the second kind. Society for Industrial and Applied Mathematics

[4] Beaney RP (1984) Positron emission tomography in the study of human tumors. Seminars in Nuclear Medicine 14(4):324–341, DOI 10.1016/S0001-2998(84)80006-9

[5] Bolomey J, Izadnegahdar A, Jofre L, Pichot C, Peronnet G, Solaimani M (1982) Microwave Diffraction Tomography for Biomedical Applications. IEEE Transactions on Microwave Theory and Techniques 30(11):1998 –2000, DOI 10.1109/TMTT.1982.1131357

[6] Born M (1999) Principles of optics: electromagnetic theory of propagation, interference and diffraction of light, 7th edn. Cambridge University Press, Cambridge, New York

[7] Brignone M, Bozza G, Randazzo A, Piana M, Pastorino M (2008) A Hybrid Approach to 3d Microwave Imaging by Using Linear Sampling and ACO. IEEE Transactions on Antennas and Propagation 56(10):3224–3232, DOI 10.1109/TAP.2008.929504

[8] Brooks RA, Chiro GD (1975) Theory of Image Reconstruction in Computed Tomography. Radiology 117(3):561–572, DOI 10.1148/117.3.561

[9] Caorsi S, Gragnani GL, Pastorino M, Zunino G (1991) Microwave imaging method using a simulated annealing approach. Microwave and Guided Wave Letters, IEEE 1(11):331–333

[10] Chew W (1995) Waves and fields in inhomogeneous media. IEEE Press, New York

[11] Collin RE, Zucker FJ (1969) Antenna theory. McGraw-Hill

[12] Crocco L, Isernia T (2001) Inverse scattering with real data: detecting and imaging homogeneous dielectric objects. Inverse Problems 17(6):1573, DOI 10.1088/0266-5611/17/6/302

[13] Dahlback R, Rubaek T, Persson M, Stake J (2012) A System for THz Imaging of Low-Contrast Targets Using the Born Approximation. IEEE Transactions on Terahertz Science and Technology 2(3):361–370, DOI 10.1109/TTHZ.2012.2189900

[14] Dembo RS, Eisenstat SC, Steihaug T (1982) Inexact Newton Methods. SIAM Journal on Numerical Analysis 19(2):400–408, DOI 10.2307/2156954

[15] Donelli M, Franceschini D, Rocca P, Massa A (2009) Three-Dimensional Microwave Imaging Problems Solved Through an Efficient Multiscaling Particle Swarm Optimization. IEEE Trans Geosci Remote Sensing 47(5):1467–1481, DOI 10.1109/TGRS.2008.2005529

[16] Donelli M, Craddock I, Gibbins D, Sarafianou M (2011) A Three-Dimensional Time Domain Microwave Imaging Method for Breast Cancer Detection Based on an Evolutionary Algorithm. Progress In Electromagnetics Research 18:179–195

[17] Dorn O, Lesselier D (2009) Level set methods for inverse scattering—some recent developments. Inverse Problems 25(12):125,001, DOI 10.1088/0266-5611/25/12/125001

[18] Eberhart R, Kennedy J (1995) A new optimizer using particle swarm theory. In: Micro Machine and Human Science, 1995. MHS '95., Proceedings of the Sixth International Symposium on, pp 39–43, DOI 10.1109/MHS.1995.494215

[19] Eisenstat SC, Walker HF (1994) Globally Convergent Inexact Newton Methods. SIAM Journal on Optimization 4(2):393–422, DOI 10.1137/0804022

[20] Eisenstat SC, Walker HF (1996) Choosing the Forcing Terms in an Inexact Newton Method. SIAM Journal on Scientific Computing 17(1):16–32, DOI 10.1137/0917003

[21] Epstein NR, Golnabi AG, Meaney PM, Paulsen KD (2012) Conformal microwave tomography using a broadband non-contacting monopole antenna array. In: 2012 IEEE International Conference on Ultra-Wideband (ICUWB), pp 192–196, DOI 10.1109/ICUWB.2012.6340462

[22] Epstein NR, Meaney PM, Paulsen KD (2013) MR-guided conformal microwave imaging for enhanced inclusion detection within irregularly shaped volumes. In: Proc. SPIE 8672, Medical Imaging 2013, p 86720H, DOI 10.1117/12.2007939

[23] Fang Q, Meaney P, Paulsen K (2004) Microwave image reconstruction of tissue property dispersion characteristics utilizing multiple-frequency information. Microwave Theory and Techniques, IEEE Transactions on 52(8):1866–1875, DOI 10.1109/TMTT.2004.832014

[24] Fhager A, Hashemzadeh P, Persson M (2006) Reconstruction quality and spectral content of an electromagnetic time-domain inversion algorithm. IEEE Trans Biomed Eng 53(8):1594–1604, DOI 10.1109/TBME.2006.878079

[25] Fhager A, Gustafsson M, Nordebo S (2012) Image reconstruction in microwave tomography using a dielectric Debye model. Biomedical Engineering, IEEE Transactions on 59(1):156–166

[26] Garnero L, Franchois A, Hugonin JP, Pichot C, Joachimowicz N (1991) Microwave imaging-complex permittivity reconstruction-by simulated annealing. Microwave Theory and Techniques, IEEE Transactions on 39(11):1801–1807

[27] Gerberich CL, Panel ORNLM, Commission UAE (1957) On the solution of a Fredholm integral equation. University of Tennessee.

[28] Golnabi AH, Meaney PM, Paulsen KD (2013) Tomographic Microwave Imaging With Incorporated Prior Spatial Information. IEEE Transactions on Microwave Theory and Techniques DOI 10.1109/TMTT.2013.2247413

[29] Gordon R, Bender R, Herman GT (1970) Algebraic Reconstruction Techniques (ART) for three-dimensional electron microscopy and X-ray photography. Journal of Theoretical Biology 29(3):471–481, DOI 10.1016/0022-5193(70)90109-8

[30] Guardiola M, Jofre L, Capdevila S, Blanch S, Romeu J (2011) 3d uwb magnitude-combined tomographic imaging for biomedical applications. algorithm validation. Radioengineering 20(2):366–372

[31] Guardiola M, Capdevila S, Romeu J, Jofre L (2012) 3-d microwave magnitude combined tomography for breast cancer detection using realistic breast models. Antennas and Wireless Propagation Letters, IEEE 11:1622–1625

[32] Gustafsson M, He S (2000) An optimization approach to two-dimensional time domain electromagnetic inverse problems. Radio Sci 35(2):525–536, DOI 200010.1029/1999RS900091

[33] Hajihashemi M, El-Shenawee M (2008) Shape Reconstruction Using the Level Set Method for Microwave Applications. IEEE Antennas and Wireless Propagation Letters 7:92–96, DOI 10.1109/LAWP.2008.920464

[34] Hansen PC (1997) Rank-deficient and discrete ill-posed problems: Numerical aspects of linear inversion. SIAM monographs on mathematical modeling and computation, SIAM, Philadelphia

[35] Hansen T, Johansen PM (2000) Inversion scheme for ground penetrating radar that takes into account the planar air-soil interface. Geoscience and Remote Sensing, IEEE Transactions on 38(1):496–506, DOI 10.1109/36.823944

[36] Hestenes MR, Stiefel E (1952) Methods of conjugate gradients for solving linear systems. National Bureau of Standards, Journal of Research 49:409–436

[37] Holland JH (1984) Genetic Algorithms and Adaptation. In: Selfridge OG, Rissland EL, Arbib MA (eds) Adaptive Control of Ill-Defined Systems, Springer US, Boston, MA, pp 317–333

[38] Holland JH, Reitman JS (1977) Cognitive systems based on adaptive algorithms. SIGART Bull 63:49–49, DOI 10.1145/1045343.1045373

[39] Huang T, Mohan A (2007) A Microparticle Swarm Optimizer for the Reconstruction of Microwave Images. IEEE Transactions on Antennas and Propagation 55(3):568–576, DOI 10.1109/TAP.2007.891545

[40] Irishina N, Dorn O, Moscoso M (2008) A level set evolution strategy in microwave imaging for early breast cancer detection. Computers & Mathematics with Applications 56(3):607–618, DOI 10.1016/j.camwa.2008.01.004

[41] Irishina N, Alvarez D, Dorn O, Moscoso M (2010) Structural level set inversion for microwave breast screening. Inverse Problems 26(3):035,015, DOI 10.1088/0266-5611/26/3/035015

[42] Joisel A, Bolomey JC (2000) Rapid microwave imaging of living tissues. In: Medical Imaging 2000, International Society for Optics and Photonics, pp 320–330

[43] Kak A (1979) Computerized tomography with X-ray, emission, and ultrasound sources. Proceedings of the IEEE 67(9):1245–1272, DOI 10.1109/PROC.1979.11440

[44] Keller JB (1969) Accuracy and Validity of the Born and Rytov Approximations. J Opt Soc Am 59(8):1003, DOI 10.1364/JOSA.59.001003

[45] Kennedy J, Eberhart RC, et al (1995) Particle swarm optimization. In: Proceedings of IEEE international conference on neural networks, vol 4, pp 1942–1948

[46] Landweber L (1951) An Iteration Formula for Fredholm Integral Equations of the First Kind. American Journal of Mathematics 73(3):615–624, DOI 10.2307/2372313

[47] Levenberg K (1944) A method for the solution of certain non-linear problems in least squares. Q Appl Math (USA) 2:164–168

[48] Marquardt DW (1963) An Algorithm for Least-Squares Estimation of Nonlinear Parameters. Journal of the Society for Industrial and Applied Mathematics 11(2):431–441

[49] Meaney PM, Demidenko E, Yagnamurthy NK, Li D, Fanning MW, Paulsen KD (2001a) A two-stage microwave image reconstruction procedure for improved internal feature extraction. Med Phys 28(11):2358–2369

[50] Meaney PM, Paulsen KD, Pogue BW, Miga MI (2001b) Microwave image reconstruction utilizing log-magnitude and unwrapped phase to improve high-contrast object recovery. Medical Imaging, IEEE Transactions on 20(2):104–116, DOI 10.1109/42.913177

[51] Mojabi P, LoVetri J (2009) Microwave Biomedical Imaging Using the Multiplicative Regularized Gauss–Newton Inversion. Antennas and Wireless Propagation Letters, IEEE 8:645–648, DOI 10.1109/LAWP.2009.2023602

[52] Mojabi P, LoVetri J, Shafai L (2011) A Multiplicative Regularized Gauss-Newton Inversion for Shape and Location Reconstruction. IEEE Transactions on Antennas and Propagation 59(12):4790–4802, DOI 10.1109/TAP.2011.2165487

[53] Natterer F, Wubbeling F (1995) A propagation-backpropagation method for ultrasound tomography. Inverse Problems 11(6):1225, DOI 10.1088/0266-5611/11/6/007

[54] Ortega JM (2000) Iterative solution of nonlinear equations in several variables. No. 30 in Classics in applied mathematics, Society for Industrial and Applied Mathematics, Philadelphia

[55] Paige CC, Saunders MA (1982) LSQR: An Algorithm for Sparse Linear Equations and Sparse Least Squares. ACM Trans Math Softw 8(1):43–71, DOI 10.1145/355984.355989

[56] Pallone MJ, Meaney PM, Paulsen KD (2012) Surface scanning through a cylindrical tank of coupling fluid for clinical microwave breast imaging exams. Medical physics 39(6):3102–3111

[57] Peronnet G, Pichot C, Bolomey JC, Jofre L, Izadnegahdar A, Szeles C, Michel Y, Guerquin-Kern JL, Gautherie M (1983) A Microwave Diffraction Tomography System for Biomedical Applications. In: Microwave Conference, 1983. 13th European, pp 529–533, DOI 10.1109/EUMA.1983.333285

[58] Phillips DL (1962) A Technique for the Numerical Solution of Certain Integral Equations of the First Kind. J ACM 9(1):84–97, DOI 10.1145/321105.321114

[59] Pichot C, Jofre L, Peronnet G, Bolomey J (1985) Active microwave imaging of inhomogeneous bodies. Antennas and Propagation, IEEE Transactions on 33(4):416–425

[60] Pozar DM (1997) Microwave engineering, 2nd edn. Wiley, New York

[61] Raichle ME (1980) Positron-Emission Tomography. In: Weiss L, Gilbert HA, Posner JB (eds) Brain Metastasis, Springer Netherlands, Dordrecht, pp 246–253

[62] Rubaek T, Meincke P (2006) Including antenna models in microwave imaging for breast cancer screening. In: Antennas and Propagation, 2006. EuCAP 2006. First European Conference on, pp 1–6

[63] Rubaek T, Meaney P, Meincke P, Paulsen K (2007) Nonlinear Microwave Imaging for Breast-Cancer Screening Using Gauss–Newton's Method and the CGLS Inversion Algorithm. Antennas and Propagation, IEEE Transactions on 55(8):2320–2331, DOI 10.1109/TAP.2007.901993

[64] Sabouni A, Noghanian S, Pistorius S (2010) A global optimization technique for microwave imaging of the inhomogeneous and dispersive breast. Electrical and Computer Engineering, Canadian Journal of 35(1):15–24

[65] Scapaticci R, Catapano I, Crocco L (2012) Wavelet-Based Adaptive Multiresolution Inversion for Quantitative Microwave Imaging of Breast Tissues. IEEE Transactions on Antennas and Propagation 60(8):3717–3726, DOI 10.1109/TAP.2012.2201083

[66] Schaffer JD (1985) Some experiments in machine learning using vector evaluated genetic algorithms. Tech. rep., Vanderbilt Univ., Nashville, TN (USA)

[67] Semnani A, Kamyab M (2007) An Enhanced Method for Inverse Scattering Problems using Fourier Series Expansion in Conjunction with FDTD and PSO. Progress In Electromagnetics Research 76:45–64, DOI 10.2528/PIER07061204

[68] Semnani A, Kamyab M (2008) Truncated Cosine Fourier Series Expansion Method for Solving 2-D Inverse Scattering Problems. Progress In Electromagnetics Research 81:73–97, DOI 10.2528/PIER07122404

[69] Shea JD, Kosmas P, Hagness SC, Van Veen BD (2010) Three-dimensional microwave imaging of realistic numerical breast phantoms via a multiple-frequency inverse scattering technique. Medical Physics 37(8):4210, DOI 10.1118/1.3443569

[70] Slaney M, Kak A, Larsen L (1984) Limitations of Imaging with First-Order Diffraction Tomography. Microwave Theory and Techniques, IEEE Transactions on 32(8):860–874

[71] Stotzka R, Wuerfel J, Mueller TO, Gemmeke H (2002) Medical imaging by ultrasound computer tomography. In: Proc. of Medical Imaging 2002: Ultrasonic Imaging and Signal Processing, pp 110–119, DOI 10.1117/12.462144

[72] Takagi A, Tsurumi Y, Ishii Y, Suzuki K, Kawana M, Kasanuki H (1999) Clinical Potential of Intravascular Ultrasound for Physiological Assessment of Coronary Stenosis Relationship Between Quantitative Ultrasound Tomography and Pressure-Derived Fractional Flow Reserve. Circulation 100(3):250–255, DOI 10.1161/01.CIR.100.3.250

[73] Tikhonov AN (1963) Solution of incorrectly formulated problems and regularization method. Doklady Akademii Nauk SSSR 151(3):501–504

[74] Winters DW, Shea JD, Kosmas P, Van Veen BD, Hagness SC (2009) Three-Dimensional Microwave Breast Imaging: Dispersive Dielectric Properties Estimation Using Patient-Specific Basis Functions. IEEE Trans Med Imaging 28(7):969–981, DOI 10.1109/TMI.2008.2008959

[75] Zhou H, Takenaka T, Johnson JE, Tanaka T (2009) A breast imaging model using microwaves and a time domain three dimensional reconstruction method. Progress In Electromagnetics Research 93:57–70

Chapter 4
Confocal Microwave Imaging

Martin O'Halloran, Dallan Byrne, Muhammad Adnan Elahi, Raquel Cruz Conceição, Edward Jones, and Martin Glavin

4.1 Introduction

Radar breast imaging (RBI) involves the illumination of the breast with a microwave pulse. If a tumour is present, backscattered radar signals are generated due to the dielectric contrast between normal and tumour tissue at microwave frequencies. These backscattered signals are processed by an RBI beamformer to identify the presence and location of any significant scatterers within the breast.

The aim of an RBI beamformer is to spatially focus the reflected electromagnetic signals to create an image of any dielectric scatterers present within the breast [23]. An effective beamforming algorithm must clearly identify the presence and location of any tumour, while simultaneously suppressing clutter due to the normal heterogeneity of breast tissue. Most RBI beamformers are extensions of the delay-and-sum (DAS) beamformer, which implements a "time-shift and sum" algorithm to estimate the backscattered energy from a particular synthetic focal

M. O'Halloran (✉) • M.A. Elahi • E. Jones • M. Glavin
Electrical and Electronic Engineering, National University of Ireland, Galway, Ireland
e-mail: martin.ohalloran@nuigalway.ie; m.elahi1@nuigalway.ie; edward.jones@nuigalway.ie; martin.glavin@nuigalway.ie

D. Byrne
Faculty of Engineering, University of Bristol, Bristol, UK

formerly at Electrical and Electronic Engineering, National University of Ireland, Galway, Ireland

Department of Electrical & Electronic Engineering, University of Bristol, Bristol, UK
e-mail: dallan.byrne@bristol.ac.uk

R.C. Conceição
Faculdade de Ciências, Instituto de Biofísica e Engenharia Biomédica, Universidade de Lisboa, Lisbon, Portugal
e-mail: raquelcruzconceicao@gmail.com

© Springer International Publishing Switzerland 2016
R.C. Conceição et al. (eds.), *An Introduction to Microwave Imaging for Breast Cancer Detection*, Biological and Medical Physics, Biomedical Engineering,
DOI 10.1007/978-3-319-27866-7_4

point within the breast [2, 14, 17, 18, 20, 34]. This synthetic focal point is then scanned throughout the breast to create a three-dimensional energy profile. Regions of high energy in the resultant breast image may suggest the presence of cancer. More sophisticated beamformers will seek to compensate for attenuation and phase effects as the microwave signals propagate to and from the synthetic focal point of interest [4, 11, 42, 43], reward the coherence of reflections from a particular synthetic focal point within the breast [8, 26, 35], and/or compensate for time-shifting or steering errors [6, 28, 56].

A number of RBI beamformers are designed based on several assumptions regarding the dielectric properties of the breast. These assumptions include the following:

- There exists a dielectric contrast between healthy breast tissue and cancerous tissue at microwave frequencies;
- The breast is primarily dielectrically homogeneous;
- The dielectric properties of the breast allow for constructive addition at the tumour site.

RBI data acquisition systems can be categorised as either monostatic or multistatic. Monostatic imaging systems transmit and receive using the same antenna, which can be physically repositioned over the exterior of the breast [15, 16, 33, 49, 53] or fixed as an element of an antenna array [28]. While monostatic systems can illuminate the breast from a range of angles, the number of acquisition positions is limited by the acceptable scanning time where a significant amount of data must be acquired relatively quickly to ensure the breast remains static during the scan [16, 29]. Fear et al. [16] reported a scanning time of 30 min to record 180 backscattered monostatic waveforms.

In multistatic systems, each element of a fixed-element array illuminates the breast in-turn while the other antennas record scattering at various angles from the transmitter boresight [9, 28, 41, 47]. Due to the spatial diversity of the receiving antennas, the multistatic approach acquires information about any scatterer present using received signals that propagate outwards via different routes. As described by Xie et al. [56], the multistatic approach can produce better imaging results when the actual aperture used in the multistatic system is close to the synthetic aperture used in the monostatic case. The term 'quasi-multistatic' refers to a variant of the multistatic approach where only a subset of the multistatic channels are used [43]. Multistatic systems incorporate static fixed-element arrays where the design goal is to populate the aperture with as many active antenna elements as possible [9, 28, 47]. The number of illuminating paths is limited by the array geometry but various scattering angles are recorded upon each transmission ensuring a significant amount of data can be acquired in a relatively short time frame. A full multistatic measurement acquisition of 1770 signals using a 60 antenna array can be performed in under 10 s, as reported in Byrne et al. [6].

Prior to image formation using beamforming, it is often necessary to remove the early-stage artefact. The artefact can consist of the transmitted pulse, skin reflections, and antenna reverberation. A wide variety of artefact removal algorithms have been proposed, and a representative selection of these algorithms is described in this chapter. Similarly, a number of the most widely used qualitative radar beamforming techniques are also examined. The chapter concludes by discussing algorithms to estimate the average dielectric properties of the breast (and consequently, the average propagation speed of the microwave signal) used to optimise the quality of the resultant microwave breast image. Note: Mathematical formulations are provided for clarity where scalars are represented as v, vectors are denoted as $\mathbf{v}$ and matrices as $\mathbf{V}$. $\mathbf{v}^T$ is the transpose of $\mathbf{v}$, and $\mathbf{v}^*$ denotes the conjugate transpose.

4.2 Artefact Removal Algorithms

The skin surrounding the breast has a thickness of generally between 1 and 3 mm [46, 50] and has significantly higher dielectric properties at microwave frequencies than the immersion medium surrounding the breast. This contrast creates a significant reflection which must be removed prior to imaging as it can mask the internal scatterers and generate significant clutter in resultant energy profiles. A number of artefact removal techniques are presented within this section.

4.2.1 Averaging Method

The averaging method [15] approximates the skin calibration signal by averaging signals across all channels. The approximated signal is then subtracted from the mth signal $\mathbf{x}_m$ as

$$s_m(n) = x_m(n) - \frac{1}{M} \sum_{i=1}^{M} x_i(n) \tag{4.1}$$

where s_m is the signal without artefact, M is the number of signals, and n denotes the sample number.

The Woody averaging [55] can be employed instead of standard averaging to compensate for channel-to-channel signal variations [13].

4.2.2 Differential Rotation

The differential rotation method uses a differential calibration system to eliminate mutual coupling and skin reflections [28]. After the first scan $\mathbf{x}_m^1$ is performed, the array is rotated by a specific angle and a second dataset is acquired $\mathbf{x}_m^2$. The signal is calibrated as

$$s_m = x_m^1 - x_m^2. \tag{4.2}$$

4.2.3 Wiener Filter

The Wiener filter algorithm [4, 11] improves on the simple average subtraction method by compensating for channel-to-channel variation in artefacts due to local variations in skin thickness, breast heterogeneity, and differences in antenna-skin distances. In this method, the artefact in each signal is estimated as a filtered combination of the signals in all other channels. The estimated artefact signal for Channel m is then subtracted from the received signal at Channel m as

$$s_m(n) = x_m(n) - q^T \mathbf{b}_{PN}(n) \tag{4.3}$$

where $\mathbf{b}_{PN}(n)$ is a vector calculated from all other channels except m, and q is the vector of filter weights. The filter weights are chosen to minimise the residual signal mean-squared error over the portion of the signal dominated by the artefact. For example, in order to remove the artefact from Channel 1, a $(2J + 1) \times 1$ vector of time samples in the mth channel is defined as

$$\mathbf{b}_m(n) = [x_m(n - J), \ldots, x_m(n), \ldots, x_m(n + J)]^T \tag{4.4}$$

where J is the number of samples on either side of nth time sample and $2J + 1$ is the length of the averaging window centred on n. The samples of $\mathbf{b}_k(n)$ for Channels 2 through N are concatenated into a vector $\mathbf{b}_{2N}(n)$ as

$$\mathbf{b}_{2N}(n) = \left[\mathbf{b}_2^T(n), \mathbf{b}_3^T(n), \ldots, \mathbf{b}_N^T(n) \right]^T. \tag{4.5}$$

The filter weight vector q is calculated as

$$q = \arg\min_q \sum_{n=n_0}^{n_0+L-1} \left| x_1(n) - q^T \mathbf{b}_{2N}(n) \right|^2 \tag{4.6}$$

where the time-window $n \in [n_0, n_0 + 1, \ldots, n_0 + L - 1]$ represents the initial portion of the signal dominated by the unwanted artefact. The solution to (4.6) is given by

$$q = \mathbf{R}^{-1}\mathbf{p} \tag{4.7a}$$

$$\mathbf{R} = \frac{1}{L} \sum_{n=n_0}^{n_0+L-1} \mathbf{b}_{2N}(n)\mathbf{b}_{2N}^T(n) \tag{4.7b}$$

$$\mathbf{p} = \frac{1}{L} \sum_{n=n_0}^{n_0+L-1} \mathbf{b}_{2N}(n) x_1^T(n). \tag{4.7c}$$

4.2.4   Root Least Squares Filter

The root least squares (RLS) filter recursively computes and updates the filter weights for the Wiener adaptation described in Sect. 4.2.3 [49]. $\mathbf{u}_m$ represents the $1 \times N$ desired signal containing N time samples at Channel m and $\hat{\mathbf{u}}(n) = \left[u_{m+1}(n), u_{m+2}(n), \ldots, u_{m+Q}(n) \right]^T$ is a $Q \times 1$ vector. The corresponding weights of the RLS filter are

$$\mathbf{w}(n) = \left[w_{m+1}(n), w_{m+2}(n), \ldots, w_{m+Q}(n) \right]. \tag{4.8}$$

The processed signal can be described as

$$s_m(n) = x_m(n) - \hat{d}_m(n) \tag{4.9}$$

where the approximated skin calibration signal is denoted by

$$\hat{d}_m(n) = \mathbf{w}^T(n)\hat{\mathbf{u}}(n). \tag{4.10}$$

Let $d_m(n)$ represent the true calibration signal and the sum of the squared error is

$$j(k) = \sum_{n=1}^{k} \lambda^{k-1} \left| d_m(n) - \hat{d}_m(n) \right|^2. \tag{4.11}$$

Expanding (4.11) results in

$$j(k) = \sum_{n=1}^{k} \lambda^{k-1} d_m^2(n) - 2\mathbf{w}^T(k) \sum_{n=1}^{k} \lambda^{k-1} \hat{\mathbf{u}}(n) d_m^T(n)$$
$$+ \mathbf{w}^T(k) \left[\sum_{n=1}^{k} \lambda^{k-1} \hat{\mathbf{u}}(n) \hat{\mathbf{u}}^T(n) \right] \mathbf{w}(k). \tag{4.12}$$

Defining:

$$a(k) = \sum_{n=1}^{k} \lambda^{k-1} d_m^2(n) \tag{4.13a}$$

$$\mathbf{B}(k) = \sum_{n=1}^{k} \lambda^{k-1} \hat{\mathbf{u}}(n) \hat{\mathbf{u}}^T(n) \tag{4.13b}$$

$$\mathbf{c}(k) = \sum_{n=1}^{k} \lambda^{k-1} \hat{\mathbf{u}}(n) d_m^T(n). \tag{4.13c}$$

Now (4.12) can be written as

$$j(k) = a(k) - 2\mathbf{w}^T(k)\mathbf{c}(k) + \mathbf{w}^T(k)\mathbf{B}(k)\mathbf{w}(k). \tag{4.14}$$

Minimisation of the above function results in the Wiener–Hopf equation

$$\mathbf{B}(k)\mathbf{w}(k) = \mathbf{c}(k) \tag{4.15}$$

$\mathbf{w}(k)$ is obtained by solving (4.15) using the matrix inversion Lemma to obtain $\mathbf{B}^{-1}$. The weight vector is calculated for each time sample, whereas in Bond's approach [4], the weight vector is calculated once for a temporal window.

4.2.5 Entropy-Based Time Windowing

The entropy-based time windowing (ETW) method uses an entropy-based time-window function to remove the skin reflections [58]. Entropy can be employed to measure the similarity of radar signals received across different channels. A larger value of entropy is obtained from the similar skin reflections in the early portion of all radar signals and conversely the tumour reflections result in much lower entropy values.

In order to compute the entropy, a probability density function along the antenna axis can be defined by normalising each received radar signal as

$$p_m[n] = \frac{\|x_m[n]\|^2}{\sum_{m=1}^{M} \|x_m[n]\|^2} \tag{4.16}$$

where $x_m[n]$ is the received signal at the mth channel and M is the total number of channels. (4.16) satisfies $p_m[n] \geq 0$ and $\sum_{m=1}^{M} p_m[n] = 1$ and can be interpreted as the energy density in the antenna domain. The general $\alpha - order$ Renyi entropy at time sample n is defined as:

$$H_\alpha[n] = \frac{1}{1-\alpha} \log \left\{ \sum_{m=1}^{M} (p_m[n])^\alpha \right\} \qquad (4.17)$$

where α is real-positive and the entropy varies from zero for certain events to log M for uniform distribution. A third-order Renyi entropy is defined for a broad class of signals [1]; therefore, $\alpha = 3$ is used in (4.17). Next, the theoretical dimension of $\left[b_1[n], b_2[n], \ldots, b_M[n] \right]$ is defined as

$$D[n] = e^{H_3^s[n]} \qquad (4.18)$$

where H_3 is the third-order Renyi entropy. The smoothed entropy is denoted as

$$H_3^s[n] = \frac{1}{K} \sum_{k=n}^{k=n+K} H_3[k] \qquad (4.19)$$

where K is the length of the smoothing window.

$D[n]$ varies between 1 and M, and the time-window function is obtained by comparing $D[n]$ to a threshold T, as in

$$W[n] = \begin{cases} 0, & \text{if } D[n] > T, \text{ where } 1 < T < M \\ 1, & \text{otherwise} \end{cases} . \qquad (4.20)$$

The time-window function $W[n]$ is applied to each radar signal to obtain skin-artefact removed signals as

$$s_m[n] = W[n]x_m[n]. \qquad (4.21)$$

4.2.6 Frequency Domain Skin-Artefact Removal

The frequency domain skin-artefact removal method [39] represents the frequency response of each received radar signal as a sum of complex exponentials, where each exponential represents a pole of the system and each pole corresponds to a specific scatterer in the view of the antenna. The skin-artefacts are removed by identifying and removing the pole corresponding to the strongest scatterers from the frequency response, assuming skin is the strongest scatterer within the signal. The frequency response of each received signal can be decomposed into its poles as

$$y(s) = \sum_{i=1}^{M} a_i e^{(\alpha_i + j\frac{4\pi}{c} R_i)s\Delta f} \qquad (4.22)$$

where M is the total number of scatterers or poles of the system, a_i is the constant coefficient, α_i is the frequency decay/growth factor, R_i is the range of the ith scatterer, and Δf is the sampling frequency. The received signals are first converted to the frequency domain using the fast Fourier transform (FFT) algorithm. These frequency domain signals are then processed using a linear system identification method to estimate the frequency model given in (4.22) [45]. The frequency domain signal is arranged in the form of a Hankel matrix in

$$H = \begin{bmatrix} y_i(1) & \cdots & y_i(L) \\ \vdots & \ddots & \vdots \\ y_i(S - L + 1) & \cdots & y_i(S) \end{bmatrix} \tag{4.23}$$

where $y_i(n)$ is nth frequency sample at Channel i and S is the total number of frequency samples. The Hankel matrix is then decomposed into the 'signal plus noise' and 'noise only' subspaces using singular value decomposition as

$$H = \begin{bmatrix} U_{sn} & U_n \end{bmatrix} \begin{bmatrix} \Sigma_{sn} & 0 \\ 0 & \Sigma_n \end{bmatrix} \begin{bmatrix} V_{sn}^* \\ V_n^* \end{bmatrix} \tag{4.24}$$

where subscript notation sn is used for 'signal-plus-noise', and n is used for noise subspace. The Akaike information criterion [10] can be used to separate the two subspaces. Removing the noise subspace, H can be approximated as

$$\tilde{H} = U_{sn} \Sigma V_{sn}^* \tag{4.25}$$

where U_{sn} is the left unitary matrix of the signal-plus-noise subspace, V_{sn}^* is the right unitary matrix of signal-plus-noise subspace, and Σ contains the dominant singular values of H in descending order and $(*)$ denotes conjugate transpose.

The approximated $\tilde{H}$ from (4.25) is further factorised using a balanced coordinate method [45]

$$\tilde{H} = \Omega \Gamma \tag{4.26}$$

where Ω is the observability and Γ is the controllability matrix defined as

$$\Omega = U_{sn} \Sigma_{sn}^{1/2} \tag{4.27a}$$

$$\Sigma = \Sigma_{sn}^{1/2} V_{sn}^*. \tag{4.27b}$$

Either Ω or Γ can be used to derive matrix A. Using Ω, A is defined as follows:

$$A = (\Omega_{-rl}^* \Omega_{-rl})^{-1} (\Omega_{-rl}^* \Omega_{-rf}) \tag{4.28}$$

where Ω_{-rl} and Ω_{-rf} are obtained by removing the last and first row of Ω. The eigenvalues of $\mathbf{A}$ are related to the range R_i and damping factor α_i of the ith scatterer as in

$$R_i = -c \frac{\phi_i}{4\pi \Delta f} \tag{4.29a}$$

$$\alpha_i = \frac{-\log |\lambda_i|}{\Delta f} \tag{4.29b}$$

where λ_i is the ith eigenvalue with phase ϕ_i. The constant coefficient a_i is denoted as

$$a_i = \frac{(\mathbf{Ce_i})(\mathbf{v_i B})}{\lambda_i^{\left(\frac{f_1}{\Delta f}\right)}} \tag{4.30}$$

where $\mathbf{C}$ is the first row of Ω, f_1 is the carrier frequency of the transmitted pulse, $\mathbf{e_i}$ are the eigenvectors of $\mathbf{A}$, and $\mathbf{v_i}$ are defined as row vectors of $\mathbf{V}$

$$\mathbf{V} = [\mathbf{e_1}, \ldots, \mathbf{e_p}]^{-1} = \begin{bmatrix} \mathbf{v_1} \\ \vdots \\ \mathbf{v_p} \end{bmatrix}. \tag{4.31}$$

$\mathbf{B}$ is denoted by

$$\mathbf{B} = (\Omega_S^* \Omega_S)^{-1} (\Omega_S^* \mathbf{y}^T) \tag{4.32}$$

where $\mathbf{y}$ is the vector of frequency samples of the backscattered signals and Ω_S is defined as

$$\Omega_S = \begin{bmatrix} \mathbf{C} \\ \mathbf{CA} \\ \vdots \\ \mathbf{CA^{S-1}} \end{bmatrix}. \tag{4.33}$$

The range, damping factor, and constant coefficients required in (4.22) can be computed from (4.29a) and (4.30).

The constant coefficients a_i are directly proportional to the amplitude of the pulse in the time domain. Considering that the amplitude of the skin reflection is quite large compared to the tumour response (in monostatic signals), a_i values over a certain threshold are removed from the frequency response of the signals. This removes the skin-related poles from the signals. The frequency response is then reconstructed using the model given in (4.22) without the dominant poles and converted back to time domain using the inverse-FFT which only contains tumour reflections.

4.2.7 Hybrid Artefact Removal

The hybrid artefact removal (HAR) method [12] combines the ETW and Wiener filter to estimate and filter the skin reflections. In this algorithm, an artefact-dominant time-window is estimated based on the entropy of radar signals and the Wiener filter is then applied to estimate the skin calibration signal for each channel, and then subtracted to reduce the skin reflections, as given in (4.3).

In order to estimate the artefact-dominant time-window, the criterion to define the time-window $W[n]$ given in (4.20) is modified. Firstly, the maxima of the function $D[n]$ are computed. The first maximum represents the point in time where signals across all channels have highest similarity and the minima indicates the maximum variation. The highly similar part of the signals is assumed to be the artefact. The time-window is then defined from the significant local minima (n_0') on the left side of the maximum to the significant local minima (m_0') on the right side of the maximum on the theoretical dimension curve.

The filter weights $\mathbf{q}$ in (4.3) are optimised over the estimated artefact-dominant time-window as

$$\mathbf{q} = \arg\min_{\mathbf{q}} \sum_{n=n_0'}^{m_0'} \left| x_m(n) - \mathbf{q}^T \mathbf{b}_{2N}(n) \right|^2 \tag{4.34}$$

where n_0' and m_0' are obtained from the entropy calculation and represent the start and the end of the artefact-dominant window, respectively. $x_m(n)$ is the target signal at Channel m and $\mathbf{b}_{2N}(n)$ is a vector calculated from signals with similar skin responses (e.g. signals received at antennas positioned at the same elevation as antenna m) excluding $x_m(n)$.

The HAR algorithm uses a subset of channels to estimate the artefact-dominant time-window for a target channel and the same set of channels is used to estimate and filter the artefact; unlike the original Wiener filter which uses all other channels except the target channel.

4.2.8 Neighbourhood-Based Skin Subtraction

The neighbourhood-based skin subtraction algorithm [37] uses a Wiener filter to estimate the skin response in a particular channel from the neighbourhood channels containing similar skin responses. The filter weights are optimised over a skin-dominant time-window estimated using an automatic window selection method. The estimated skin response is then subtracted from a target channel to obtain the skin subtracted signal.

The radar signals are first processed through a skin-dominant window selection algorithm. The maxima and minima of the signal are determined by computing the first- and second-order derivatives of the signal. The significant local maxima are assumed to be the part of the skin because the skin response is typically several orders of magnitude larger than the late-time response in the signal. The significant local maxima are determined with a threshold cut-off based on the antenna characteristics and is known as a priori. All adjacent peaks are considered to be the part of the artefact except the first and last peak. The start of the skin-dominant window, n_0', is set equal to the time step corresponding to the trough that precedes the first significant peak. The end of the skin-dominant window m_0' is defined as the time step corresponding to the trough that follows the last significant peak.

After the estimation of the skin-dominant window for each channel, the next step is to define the neighbourhood channels. The neighbourhood channels are defined based on the antenna proximity and the cross-correlation between each of the recorded signals [37] and $x_m(n)$. An antenna is considered to be in the neighbourhood of the target antenna if it is within twice the half-energy beamwidth (HEB) of the target antenna in the horizontal direction and less than HEB minus 1 cm in the vertical direction. The difference in the proximity criterion for the horizontal and vertical direction is due to the significant variation of the breast shape in the vertical direction compared to the horizontal direction. Hence a subgroup of neighbouring antennas is defined based on the proximity criterion. The reflections from each neighbouring antenna are then cross-correlated with the reflection from the target antenna to validate the similarity of reflections. A threshold is then used to validate the similarity of reflections from antennas included in the neighbourhood to the target antenna. Any antenna not meeting the similarity criteria is excluded. The similarity criteria ensures that antennas selected in the neighbourhood have enough similarity to provide an accurate estimate of the skin response in the target channel.

The skin response for each channel is estimated from neighbouring channels and subtracted from the target channel as

$$s_m(n) = x_m(n) - q^T b_{\text{neighbours},m}(n) \tag{4.35}$$

where m is the target antenna, $b_{\text{neighbours},m}(n)$ represent the collection of signals from the neighbouring antennas of m, and q are the filter weights calculated as

$$q = \arg\min_q \sum_{n=n_0'}^{m_0'} \left| x_m(n) - q^T b_{\text{neighbours},m}(n) \right|^2 \tag{4.36}$$

with the solution given in (4.7a).

4.2.9 Independent Component Analysis Artefact Removal Algorithm

The independent component analysis (ICA) method [8] is used to estimate and remove the skin calibration signal by increasing the amount of measurement data. The measurement data is increased by repositioning the antenna array. A skin calibration signal $\bar{\mathbf{x}}_{i,j,k}$ is derived from scan k for each measurement signal $\mathbf{x}_{i,j,k}$ and the calibrated signal is denoted as

$$\mathbf{s}_{i,j,k} = \mathbf{x}_{i,j,k} - \bar{\mathbf{x}}_{i,j,k}. \tag{4.37}$$

At each repositioning of the fixed-element array, the geometrical array arrangement results in consistent antenna coupling effects and skin propagation among corresponding transmit–receive (TX-RX) signals, i.e. between $\left[\mathbf{x}_{i,j,1}, \mathbf{x}_{i,j,2}, \ldots, \mathbf{x}_{i,j,K}\right]$. Other TX-RX signals from intra-scan data may experience similar coupling and skin scattering, particularly if they are geometrically similar to another bistatic arrangement in the array, e.g. a pair of bistatic paths whose antennas exhibit similar relative polarisations and phase centres are equidistant. These intra-scan signals and the corresponding signal pairs present in the other $K - 1$ scan positions are collated into a group. Grouped signals which vary significantly from $\mathbf{x}_{i,j,k}$ are eliminated through a normalised cross-correlation thresholding factor, denoted by xC, resulting in a group matrix $\mathbf{M}_{i,j,k}$. A mathematical outline of the normalised cross-correlation is well documented in [38]. A threshold of $xC >= 0.97$ is used throughout and was found to offer robust skin calibration by ensuring that a sufficient number of signals are present to form a group. In contrast to other proposed filtering methods [3, 43, 49], no time windowing is used in this study. Columns of $\mathbf{M}_{i,j,k}$ represent m matched channels and rows contain N samples.

Data can be pre-whitened using methods such as principal component analysis so that components are uncorrelated and their variances equal unity [25]. Let $\hat{\mathbf{M}}$ represent the whitened form of $\mathbf{M}_{i,j,k}$, where the subscript notation is omitted temporarily for simplicity. ICA then attempts to separate source signals from unwanted interference and noise [22]. The method assumes that a single measurement is a linear mixture of non-Gaussian sources and independent components are obtained by searching for a linear combination of the signal data which maximises this non-Gaussianity. The linear ICA model can be denoted by

$$\hat{\mathbf{M}} = \mathbf{AS} \tag{4.38}$$

where the N rows of $\mathbf{S}$ describe the p sources and $\mathbf{A}$ is the $N \times p$ mixing matrix. The $p < m$ independent components are calculated by

$$\mathbf{S} = \mathbf{W}^T \hat{\mathbf{M}} \tag{4.39}$$

where $\mathbf{W}$ is the inverse of $\mathbf{A}$ and is estimated through a maximisation of the non-Gaussianity of the independent components. Calculation of $\mathbf{W}$ is obtained using the FastICA method [24, 25] with the following fixed point update algorithm:

Choose random initial weight vector $\mathbf{w}$.

Repeat until convergence:

1. Let $\mathbf{w}^+ = E\{\hat{\mathbf{m}}, g(\mathbf{w}^T\hat{\mathbf{m}})\} - E\{\hat{\mathbf{m}}, g^{'}(\mathbf{w}^T\hat{\mathbf{m}})\}\mathbf{w}$

2. Let $\mathbf{w} = \frac{\mathbf{w}^+}{\|\mathbf{w}^+\|}$

where $\mathbf{w}$ is a column of $\mathbf{W}$, $\hat{\mathbf{m}}$ is the vectorised representation of $\hat{\mathbf{M}}$, E is the expected value, and g is the derivative of the contrast function which is given in [24]. To prevent different vectors from converging to the same maxima, the outputs $\mathbf{w}_1^T \hat{\mathbf{m}}, \ldots, \mathbf{w}_p^T \hat{\mathbf{m}}$ must be uncorrelated after every iteration using a deflation scheme based on a Gram-Schmidt-like method, described in detail in [24].

The number of independent components p is determined by eliminating principal components whose corresponding eigenvalues are below a specified threshold [22, 25]. The resulting columns of $\mathbf{S}$ are unordered. A cross-correlation is performed against the mean of $\hat{\mathbf{M}}$ to extract the independent components which best represent the undesirable reflections, resulting in a skin calibration signal $\bar{\mathbf{x}}_{i,j,k}$.

4.3 Data Independent Beamformers

4.3.1 Delay-and-Sum Beamformer

The DAS synthetic focusing technique [2] is the basis for the vast majority of breast imaging algorithms and is described here for a monostatic antenna configuration, as in [18, 21, 33]. Considering a location r_0 within the breast, the time delay to and from each antenna to r_0 is calculated by assuming both a straight-line propagation path and the average electromagnetic wave speed through the breast. The average propagation speed is established based on an estimate of the average dielectric properties of the breast at the centre frequency of the transmitted wideband pulse. These time delays are then used to isolate the response from the synthetic focal point r_0 within each monostatic radar signal. The responses are summed, squared, and integrated over a predefined time-window T_{Win} to produce a qualitative intensity value associated with the point r_0, as

$$I(r_0) = \int_0^{T_{\mathrm{Win}}} \left[\sum_{m=1}^M x_m\big(\tau_m(r_0)\big) \right]^2 dt \qquad (4.40)$$

where $\mathbf{x}_m$ is the backscattered radar signal recorded at the mth antenna and $\tau_m(r_0)$ is defined as follows:

$$\tau_m(r_0) = \frac{d_m}{v} f_s \qquad (4.41)$$

where d_m is the round-trip distance from the mth transmitting antenna to the point r_0, f_s is the sampling frequency, and v is the average speed of propagation in breast tissue and is defined as

$$v = \frac{c}{\sqrt{\epsilon_r}} \qquad (4.42)$$

where ϵ_r is the relative permittivity of breast tissue at the centre frequency of the input signal and c is the speed of light in a vacuum. The synthetic focal point r_0 is then scanned throughout the breast and an image of the backscattered energy is created.

In order to utilise all possible transmit–receive combinations in the antenna array, Nilavalan et al. [40] adapt the DAS beamformer for a multistatic array configuration. In this case, each antenna transmits a UWB pulse in-turn and the reflected signals are recorded at all antennas. The monostatic signals and reciprocal signals are excluded (as they would not provide any additional information), leaving $M(M-1)/2$ radar signals. An additional weighting factor, w, is used to compensate for path-dependent attenuation. The intensity at a particular focal point within the breast image is as follows:

$$I(r_0) = \int_0^{T_{\text{Win}}} \left[\sum_{m=1}^{M(M-1)/2} w_i x_m \left(\tau_m(r_0) \right) \right]^2 dt. \tag{4.43}$$

Nilavalan suggests that the increased number of observations offered by the multistatic approach should provide for increased clutter suppression in the resultant breast images.

4.3.2 Delay-Multiply-And-Sum Beamformer

The delay-multiply-and-sum (DMAS) beamformer [35] adds an additional processing step to the original DAS beamformer. This step involves the pair-multiplication of the time-aligned radar signals prior to summation.

The process effectively rewards the 'coherence' or 'correlation' of the reflections from the tumour location. Reflections with high levels of coherence across the multistatic channels are enhanced through the pair-multiplication process. Since reflections from tumours should exhibit a high level of coherence, the tumours (or any dielectric scatterer) will appear brighter relative to the background clutter in the resultant image.

Using the DMAS beamformer, the intensity at a particular focal point r_0 within the breast is calculated as follows:

$$I(r_0) = \int_0^{T_{\text{Win}}} \left[\sum_{m=1}^{M-1} \sum_{j=(m+1)}^{M} x_m \left(\tau_m(r_0) \right) \odot x_j \left(\tau_j(r_0) \right) \right]^2 dt \tag{4.44}$$

where $\odot$ represents an element-by-element multiplication, and m and j describe exclusive signal indices. Again, the focal point is scanned throughout the breast to create a scattering profile of the breast.

4.3.3 Coherence Weighted Beamformers

The improved-delay-and-sum beamformer (IDAS) method [26] extends the traditional DAS beamformer by introducing an additional weighting factor called the 'Quality Factor' (QF). In a similar fashion to the DMAS beamformer, the QF can be interpreted as a measure of the coherence of the reflected radar signals corresponding to a particular synthetic focal point r_0 within the breast. As mentioned in the previous section, high coherence should correspond to reflections from a tumour (or any other dielectric scatterer) within the breast.

The QF is calculated in a three part process:

1. For a particular focal point, a curve of energy collected versus number of signals summed is created;
2. Next, this curve is rescaled by normalising it to the standard deviation of energy, $sigma_e$, for all radar signals used in the summation;
3. Finally, a second-order polynomial is fitted (using least square) to the curve of coherent energy collection ($y = ax^2 + bx + c$).

If the reflections are perfectly time-aligned, then the curve of energy collection should form a quadratic curve. Therefore, the QF is assigned the value a from the second-order polynomial equation. Finally, the QF is incorporated into the beamformer equation as follows:

$$I(r_0) = QF(r_0) \int_0^{T_{\text{Win}}} \left[\sum_{m=1}^{M} x_m\Big(\tau_m(r_0)\Big) \right]^2 dt \tag{4.45}$$

where $QF(r_0)$ is the quality factor at the focal point r_0.

In a separate study, Klemm et al. replace the quality factor with a "Coherence Factor" (CF) [28]. Using a concept adapted from ultrasound signal processing, a CF value for a particular focal point is calculated to minimise the variance across the channels as follows:

$$CF(r_0) = \frac{\left| \sum_{m=1}^{M} x_m\Big(\tau_m(r_0)\Big) \right|^2}{\sum_{m=1}^{M} \left| x_m\Big(\tau_m(r_0)\Big) \right|^2}. \tag{4.46}$$

The CF is then incorporated into the beamformer along with an additional weighting factor $w_m(r_0)$

$$I(r_0) = CF(r_0) \odot \int_0^{T_{\text{Win}}} \left[\sum_{m=1}^{M} w_m(r_0) x_m\Big(\tau_m(r_0)\Big) \right]^2 dt \tag{4.47}$$

$w_m(r_0)$ is a location dependent weight which aims to compensate for path-dependent attenuation.

Yin et al. [57] derive an alternative weighting factor system which rewards the correlation between monostatic time shifted signals at r_0. The normalised cross-correlation between delayed signals from neighbouring antennas is computed and multiplied to obtain a scalar weighting factor. The normalised correlation factor is denoted as

$$c_{ij}(r_0) = \mathrm{xcorr}\left(x_n\left(\tau_n(r_0)\right), x_m\left(\tau_m(r_0)\right)\right) \qquad (4.48)$$

for neighbouring antenna indices m and n. These values are normalised, sorted in descending order and the top $(M-1)/2$ are retained as

$$\hat{\mathbf{c}}(r_0) = \left[\hat{c}_1, \hat{c}_2, \ldots, \hat{c}_{(M-1)/2}\right] \qquad (4.49)$$

where $\hat{c}_1 > \hat{c}_2 > \cdots > \hat{c}_{(M-1)/2}$. The weighting can be described as

$$w(r_0) = \prod_{i=1}^{(M-1)/2} c_i(r_0). \qquad (4.50)$$

The energy at r_0 can be denoted as

$$I(r_0) = \int_0^{T_{\mathrm{Win}}} \left[w(r_0) \sum_{m=1}^{M} x_m(\tau_m(r_0))\right]^2 dt. \qquad (4.51)$$

4.3.4 Microwave Space-Time Beamformer

The microwave space-time (MIST) beamformer [3] aims to compensate for frequency-dependent propagation effects to better spatially focus the backscattered signals, providing an improved image of the breast.

The MIST method first time-shifts the received radar signals to align the reflections from a particular focal point within the breast (in a similar fashion to the DAS beamformer previously described). The time-aligned signals are then processed by a bank of finite-impulse response (FIR) filters. These filters are designed to compensate for path-dependent attenuation and phase effects in the reflected radar signals. This ensures that reflections from tumours deep within the breast are as strong following filtering as reflections from shallower dielectric scatterers.

For each multistatic channel m, the FIR filter of length L can be denoted by $\mathbf{w}_m = \left[w_{m0}, w_{m1}, \ldots, w_{m(L-1)}\right]^T$. The frequency response of each filter is given by

$$W_m(\omega) = \sum_{l=0}^{L-1} \omega_{ml} e^{-j\omega l T_s} = \mathbf{w}_m^T \mathbf{d}(\omega) \qquad (4.52)$$

where $\mathbf{d}(\omega) = \left[1, e^{-j\omega T_s}, \ldots, e^{-j\omega(L-1)T_s}\right]^T$ and T_s is the sampling interval. Assuming the time-alignment step has been completed, the filter coefficients must satisfy the following equation:

$$\sum_{m=0}^{M} S_m(r_0, \omega)\mathbf{w}_m^T\mathbf{d}(\omega) \approx e^{-j\omega T_s(L-1)/2} \qquad (4.53)$$

where $S_m(r_0, \omega)$ is a model of the round-trip channel as the input signal propagates through the mth channel to the target and is defined as follows:

$$S_m(r_0, \omega) = \left[\frac{1}{\sqrt{d_m}}e^{-\alpha(\omega)d_m}e^{-j\beta(\omega)d_m}\right], \qquad (4.54)$$

where d_m is the distance between the transmitting antenna, focal point, and receiver, $\alpha(\omega)$ is the frequency-dependent attenuation factor, and $\beta(\omega)$ is the frequency-dependent phase constant.

O'Halloran et al. [43] extend the MIST technique for an application with multistatic breast radar data.

4.4 Adaptive Beamforming

In contrast to the data independent methods detailed above, the data adaptive beamformer derives a weighting factor from the second-order statistics, e.g. variance, of the measured signal data to reduce the levels of noise and clutter in the scattering profiles. Data adaptive beamformers involve first time-aligning the signals (in a similar fashion to the DAS beamformer), and then further processing the signals to achieve unit gain from a desired direction, while suppressing signals of the same frequency from all other directions [51]. In this section, a number of data adaptive algorithms that have been applied to the RBI problem are detailed.

4.4.1 Minimum Variance Capon beamformer

Data adaptive methods applied to breast imaging are based on the minimum variance distortionless response (MVDR) or Capon method which calculates beamformer weights to minimise the variance of the system while limiting the beamformer response to maintain a distortionless response (i.e. achieving unit gain towards the desired signal component [23]).

The complex signal data at snapshot k can be denoted as

$$\mathbf{x}(k) = \mathbf{a}s(k) + \mathbf{v}(k) \qquad (4.55)$$

where $\mathbf{a}$ is the array response vector to the arriving wavefront, $s(k)$ is the desired signal, and $\mathbf{v}$ describes noise. The beamformer output is described as

$$y(k) = \mathbf{w}^H \mathbf{x}(k) \tag{4.56}$$

and the Capon beamforming problem is defined by the following minimisation problem:

$$\min_{w} \mathbf{w}^* \mathbf{R}_{\mathbf{x}} \mathbf{w} \tag{4.57a}$$

$$\mathbf{w}^* \mathbf{a} = 1 \tag{4.57b}$$

where $\mathbf{w}^* \mathbf{a} = 1$ is the response to the desired signal component. MVDR-derived beamformers typically model statistical noise power using a covariance matrix containing only noise and interference. However, in many communications and radar imaging applications, such data are unavailable and the sample covariance matrix is constructed from received data snapshots in lieu of the noise covariance matrix. $\mathbf{R}_{\mathbf{x}}$ represents the covariance matrix, equated as $E\{\bar{\mathbf{x}}\bar{\mathbf{x}}^H\}$, and is calculated for this application as a sample covariance matrix across K predefined snapshots as

$$\mathbf{R}_{\mathbf{x}} = \frac{1}{K} \sum_{n=k-K+1}^{k} \mathbf{x}(n)\mathbf{x}^H(n). \tag{4.58}$$

MVDR methods are known to outperform data-independent beamformers providing that self-nulling does not occur, i.e. the desired signal component is absent from the snapshots used to construct the sample covariance matrix $\mathbf{R}_{\mathbf{x}}$ which is not the case for breast radar systems. Robustness to self-nulling and non-ideal propagation characteristics is added by assuming that the true steering vector ($\mathbf{a}$) is constrained by the following L2 norm (squared difference) on the uncertainty set, as

$$||\Delta\mathbf{a}|| < \epsilon \tag{4.59}$$

where $\Delta\mathbf{a} = \hat{\mathbf{a}} - \mathbf{a}$, $\hat{\mathbf{a}} = \mathbf{1}_M$ is the assumed plane-wave steering vector and ϵ is an empirically derived upper bound on the uncertainty. The Robust Capon or Robust MVDR problem can be written as

$$\text{minimise}_{\mathbf{w}} \quad \mathbf{w}^* \mathbf{R}_{\mathbf{x}} \mathbf{w}$$

$$\text{subject to} \quad |\mathbf{w}^* \mathbf{a}| \geq 1, \tag{4.60}$$

$$\forall ||\Delta\mathbf{a}|| \leq \epsilon$$

and the weights calculated as

$$\mathbf{w} = \frac{\mathbf{R}_{\mathbf{x}}^{-1}\mathbf{a}}{\mathbf{a}^*\mathbf{R}_{\mathbf{x}}^{-1}\mathbf{a}}. \tag{4.61}$$

This is the equivalent of the quadratic problem defined by Li and Stoica [31] and can be solved using Lagrange multipliers [31, 36] or reformulated for interior point methods [52].

4.4.2 Multistatic Adaptive Microwave Imaging

The multistatic adaptive microwave imaging (MAMI) algorithm by Xie et al. [56] involves applying the MVDR in two stages in the time domain. The MVDR output in the first stage provides M desired waveform estimates. However, the steering vector may still be imprecise and a second stage is required to calculate a waveform output for the complete system. The pre-steered (to r_0) input vector for transmitter i is described by

$$\mathbf{x}_i(n) = [x_1(n), x_2(n), \ldots, x_M(n)]^T, \qquad \mathbf{x}_i(n) \in \mathbf{R}^{M \times 1} \tag{4.62}$$

and the sample covariance matrix $\mathbf{R}_X(n)$ is calculated as

$$\mathbf{R}_X(n) = \frac{1}{M} \mathbf{X}(n) \mathbf{X}^T(n) \tag{4.63}$$

where

$$\mathbf{X}(n) = [\mathbf{x}_1(n), \mathbf{x}_2(n), \ldots, \mathbf{x}_M(n)] \qquad \mathbf{X}(n) \in \mathbf{R}^{M \times M}. \tag{4.64}$$

The Stage 1 waveform estimate is written as

$$\mathbf{s}(n) = [\mathbf{w}_{S1}^T(n) \mathbf{X}(n)]^T. \tag{4.65}$$

After Stage 1, the steering vector calculation is improved; however, it may be imprecise and require a second application of the RCB algorithm. The output of (4.65) is used as the input argument to the RCB method in Stage 2, where the sample covariance matrix is now determined by

$$\mathbf{R}_S = \frac{1}{N} \sum_{t=0}^{T} \mathbf{s}(n) \mathbf{s}^T(n). \tag{4.66}$$

All other steps are calculated as in Stage 1 and the energy is calculated with

$$I(r_0) = \int_0^{T_{\text{Win}}} \left[\mathbf{w}_{S2}(n) \mathbf{s}(n) \right]^2 dn. \tag{4.67}$$

Klemm et al. [27] highlight that the MAMI technique works well with measurement radar data obtained from 60-antenna elements multistatic radar array.

4.4.3 Transmitter-Grouping Robust Capon Beamformer

The transmitter-grouping RCB (TG-RCB) [7] applies the RCB beamformer to
different subgroups of multistatic signals, based on the associated transmitting
antenna. In a group with N time samples and Tx transmitting antenna elements,
the pre-steered (to r_0) input signals for the ith transmitter are organised as follows:

$$\mathbf{x}_i(n) = [x_1(n), x_2(n), \dots, x_{RX}(n)]^T. \tag{4.68}$$

The Robust Capon beamformer (RCB) problem can then be described with (4.60),
and the sample covariance matrix is defined as

$$\mathbf{R}_i = \frac{1}{T_{\text{Win}}} \sum_{t=0}^{T_{\text{Win}}} \mathbf{x}_i(n)\mathbf{x}_i^T(n). \tag{4.69}$$

The energy at a specific voxel can then be calculated over the window T_{win} as

$$I(r_0) = \int_0^{T_{\text{Win}}} \left[\sum_{i=1}^{Tx} [\mathbf{w}_i^T \cdot \mathbf{x}_i(n)] \right]^2 dn. \tag{4.70}$$

4.4.4 Wideband Time-Domain Adaptive Beamforming

Both the MAMI and TG-RCB formulations apply a narrowband beamformer
to wideband time-domain data. Byrne and Craddock [6] adapt a time-domain
stacked–tapped filter to calculate MVDR weights across the entire frequency band.
A calculated permittivity estimate [48] of the breast tissue is used to pre-steer and
aid in the equalisation process [4, 44]. The pre-steered (to r_0) equalised input at the
jth receiver $\mathbf{x}_j$ is converted to stacked-tap for a tapped-delay line filter architecture

$$\dot{\mathbf{x}}_j(n) = \left[\mathbf{x}_{j1}^T(n), \mathbf{x}_{j2}^T(n), \dots, \mathbf{x}_{jL}^T(n) \right]^T \tag{4.71}$$

where $\mathbf{x}_{jl}(n)$ denotes the $M \times 1$ array signal data after the $(l-1)$ tap obtained at
snapshot n. The minimisation problem is reformulated for an L-tap FIR filter across
N_f frequencies as

$$\min_{\mathbf{w}} \mathbf{w}^* \mathbf{R}_{\dot{x}} \mathbf{w}$$

$$\text{subject to} \quad \Re\left(h_{DS}(f_k) e^{j\omega_k \bar{T}} \right) \geq 1 \tag{4.72}$$

$$\forall \|\Delta \mathbf{a}(f_k)\| \leq \epsilon, \quad k = [1, \dots, N_f].$$

where $\mathbf{w}$ is the stacked-tap representation of $\mathbf{W}$, with W_{ml} denoting the specific filter weight for the mth signal after the $l-1$ tap. $\Re$ is the real operator, $\bar{T} = (L-1)\Delta T/2$ is the filter delay, and the beamformer response towards the desired signal $h_{DS}(f_k)$ is denoted as

$$h_{DS}(f_k) = \mathbf{w}^T(\mathbf{d}(f_k) \otimes \mathbf{1}_M) + \mathbf{w}^T(\mathbf{d}(f_k) \otimes \Delta\mathbf{a}(f_k)) \tag{4.73}$$

where $\mathbf{d}(f_k) = [1, e^{-j\omega_k\Delta T}, \ldots, e^{-j\omega_k(L-1)\Delta T}]^T$, ΔT is the time-resolution and $\otimes$ represents the Kronecker product. The sample covariance matrix is calculated as

$$\mathbf{R}_{\dot{\mathbf{x}}_j} = \frac{1}{K} \sum_{l=n-K+1}^{n} \dot{\mathbf{x}}_j(l)\dot{\mathbf{x}}_j^*(l) \tag{4.74}$$

where $K \leq T$ represents the snapshot window length.

The desired waveform for each receiver is denoted as

$$s_j(n) = \mathbf{w}^T \dot{\mathbf{x}}_j(n). \tag{4.75}$$

This process is repeated to obtain J waveform approximations

$$\mathbf{s}(n) = [s_1(n), s_2(n), \ldots, s_J(n)]^T. \tag{4.76}$$

The beamforming process is repeated to find weights $\mathbf{w}_S$ which satisfy (4.72) and result in the output waveform estimate

$$y(n) = \mathbf{w}_S^T \mathbf{s}(n) \tag{4.77}$$

and the voxel scattering energy as

$$I(r_0) = \int_0^{T_{\mathrm{Win}}} \left[y(n)\right]^2 dn. \tag{4.78}$$

4.5 Path Dielectric Estimation Techniques

Radar beamforming techniques assume the wave propagation speed in order to appropriately focus to r_0. The speed is calculated as in (4.42) where the permittivity of the media is typically assumed to represent fatty tissue [4, 15, 27, 33]; however, a study by Lazebnik et al. [30] shows that the dielectric properties of dense breast tissue can be significantly higher. A-priori knowledge of the breast density, obtained from an inverse reconstruction, can also been used to estimate the wave propagation speed [54, 57].

4.5.1 Multiple Signal Classification Time-Of-Flight

Sarafianou et al. [48] use a time-of-flight (TOF) method to determine the speed
of the signals travelling through the breast, between transmitter and receiver. This
extension of the MUltiple SIgnal Classification (MUSIC) approach [32] is adapted
for a 31 element breast radar measurement array. $x_m(f)$, the mth recorded signal in
the frequency domain, is divided into N segments as

$$\mathbf{x}_m(k) = [x_m(k), x_m(k+1), \dots, x_m(k+L-1)]^T \tag{4.79}$$

where L is the predefined length of the sliding window. The auto-correlation matrix
is defined by

$$\mathbf{R}_{\mathbf{x}_m} = \frac{1}{N} \sum_{k=0}^{N-1} \mathbf{x}_m(k)\mathbf{x}_m^*(k). \tag{4.80}$$

The number of multipath signals L_p can be determined using the minimum length
descriptor criterion [32] and the MUSIC spectrum can be computed as

$$S(\tau) = \frac{1}{\sum_{k=L_p}^{L-1} \frac{1}{\lambda_k} |\mathbf{q}_k^H \mathbf{v}(\tau)|} \tag{4.81}$$

where τ is the propagation delay, $\lambda_k \in [\lambda_0, \lambda_1, \dots, \lambda_{L-1}]$ and $\lambda_0 > \lambda_1 > \cdots >$
λ_{L-1} are the eigenvalues of $\mathbf{R}_{\mathbf{x}_m}$ and $\mathbf{q}_k$ is the corresponding kth eigenvector.
$\mathbf{v}(\tau) = [1, e^{-j2\pi \Delta f \tau}, \dots, e^{-j2\pi(L-1)\Delta f \tau}]^T$ and Δf is the frequency resolution. The
TOF through the breast is determined by the first peak in $S(\tau)$ and the speed can be
calculated using the known distance between transmitter and receiver.

4.5.2 Transmission Coefficient Method

An alternative approach proposed by Bourqui et al. [5] estimates the transmission
coefficients, specifically the attenuation and phase constants, of the channel using
a bistatic configuration. Two Cassiopeia antennas are used as the transmit–receive
pair and both elements are placed directly onto the skin and a significantly lossy
immersion liquid (2 % Saline) is used to mitigate waves travelling around the breast.
The method was evaluated with a number of volunteer patients with suspected
primarily fatty breasts. The electric field within tissue at a distance d from the
transmitter is defined as

$$E_d(f) = E_0(f)e^{-\alpha(f)d}e^{-j\beta(f)d} \tag{4.82}$$

where E_0 is the incident field, α is the attenuation constant, and β is the phase constant at frequency f. Two measurements are taken (a and b) through the breast where the channel distance is varied and the propagation coefficients can be calculated as

$$\alpha(f) = \frac{-\log\left(|S_{21}^a(f)|/|S_{21}^b(f)|\right)(D^a)/D^b)}{D^a - D^b} \tag{4.83}$$

$$\beta(f) = -\frac{\phi^a(f) - \phi^b(f)}{D^a - D^b} \tag{4.84}$$

where D describes the distance between the antennas, $D^a \neq D^b$, $|S_{21}|$ is the magnitude of the transmission coefficient, and ϕ is the phase. The permittivity $\epsilon_R(f)$ and conductivity $\sigma(f)$ are derived by

$$\epsilon_R(f) = \frac{\beta(f)^2 - \alpha(f)^2}{(2\pi f)^2 \mu_0 \epsilon_0} \tag{4.85}$$

$$\sigma(f) = \frac{2\alpha(f)\beta(f)}{2\pi f \mu_0}. \tag{4.86}$$

4.5.3 Optimisation-Based Propagation Technique

Guo et al. [19] determine the position-dependent dielectric constants based on a certain objective equation. Beamformed energy values within the domain D are calculated and collated in subgrids to form the cost function. The technique is robust in brain imaging scenarios—with and without the presence of electromagnetic scattering—when there is internal bleeding. The technique is formulated as an optimisation problem

$$\epsilon_R^{opt} = \arg\max_{\epsilon_R}\left\{\frac{\max(\mathbf{G})}{||\mathbf{G}||_1 - \max(\mathbf{G})}\right\} \tag{4.87}$$

$\Omega_k \in D$ defines the subgrid and ϵ_R^{opt} is the optimal relative permittivity to use. $\mathbf{G}$ can be derived from

$$\mathbf{G} = \left[G_1^{sub}, G_2^{sub}, \dots, G_K^{sub}\right] \tag{4.88}$$

$$G_k^{sub} = \frac{\sum_r I(r)}{\sum_r \chi_D(r)}, \quad r \in \Omega_k \tag{4.89}$$

where $\chi_D(\cdot)$ is the indicator function of D, k is the subgrid index which pertain to the cells within $\Omega_k \in D$. The objective function in (4.87) is minimised until convergence or a maximum of 50 iterations.

References

[1] Baraniuk RG, Flandrin P, Janssen AJ, Michel OJ (2001) Measuring time-frequency information content using the Rényi entropies. Information Theory, IEEE Transactions on 47(4):1391–1409

[2] Benjamin R (1996) Synthetic, post-reception focusing in near-field radar. In: The Detection of Abandoned Land Mines: A Humanitarian Imperative Seeking a Technical Solution, EUREL International Conference on (Conf. Publ. No. 431), pp 133–137, DOI 10.1049/cp:19961095

[3] Bond E, Li X, Hagness S, Van Veen B (2003a) Microwave imaging via space-time beamforming for early detection of breast cancer. IEEE Transactions on Antennas and Propagation 51(8):1690–1705

[4] Bond EJ, Li X, Hagness SC, Van Veen BD (2003b) Microwave imaging via space-time beamforming for early detection of breast cancer. Antennas and Propagation, IEEE Transactions on 51(8):1690–1705

[5] Bourqui J, Garrett J, Fear E (2012) Measurement and Analysis of Microwave Frequency Signals Transmitted Through the Breast. Journal of Biomedical Imaging 2012:1:1–1:1, DOI 10.1155/2012/562563

[6] Byrne D, Craddock I (2015) Time-Domain Wideband Adaptive Beamforming for Radar Breast Imaging. IEEE Transactions on Antennas and Propagation 63(4):1725–1735, DOI 10.1109/TAP.2015.2398125

[7] Byrne D, O'Halloran M, Jones E, Glavin M (2010) Transmitter-grouping robust capon beamforming for breast cancer detection. Progress In Electromagnetics Research 108:401–416

[8] Byrne D, Sarafianou M, Craddock I (In Press) Compound radar approach for breast imaging. Biomedical Engineering, IEEE Transactions on

[9] Craddock I, Nilavalan R, Leendertz J, Preece A, Benjamin R (2005) Experimental investigation of real aperture synthetically organised radar for breast cancer detection. In: Antennas and Propagation Society International Symposium, 2005 IEEE, vol 1, pp 179–182

[10] Cuomo KM, Piou JE, Mayhan JT (1997) Ultra-wideband coherent processing. Lincoln Laboratory Journal 10(2):203–222

[11] Davis SK, Bond EJ, Hagness S, Van Veen B (2003) Microwave imaging via space-time beamforming for early detection of breast cancer: Beamformer design in the frequency domain. Journal of Electromagnetic Waves and Applications 17(2):357–381

[12] Elahi MA, Shahzad A, Glavin M, Jones E, O'Halloran M (2014) Hybrid Artifact Removal for Confocal Microwave Breast Imaging. IEEE Antennas and Wireless Propagation Letters 13:149–152, DOI 10.1109/LAWP.2014.2298975

[13] Fear E, Sill J (2003) Preliminary investigations of tissue sensing adaptive radar for breast tumor detection. In: Engineering in Medicine and Biology Society, 2003. Proceedings of the 25th Annual International Conference of the IEEE, IEEE, vol 4, pp 3787–3790

[14] Fear E, Stuchly M (2000) Microwave detection of breast cancer. Microwave Theory and Techniques, IEEE Transactions on 48(11):1854–1863

[15] Fear E, Li X, Hagness S, Stuchly M (2002a) Confocal microwave imaging for breast cancer detection: Localization of tumors in three dimensions. IEEE Transactions on Biomedical Engineering 49(8):812–822

[16] Fear E, Bourqui J, Curtis C, Mew D, Docktor B, Romano C (2013) Microwave Breast Imaging With a Monostatic Radar-Based System: A Study of Application to Patients. IEEE Transactions on Microwave Theory and Techniques 61(5):2119–2128, DOI 10.1109/TMTT.2013.2255884

[17] Fear EC, Stuchly MA (1999) Microwave system for breast tumor detection. Microwave and Guided Wave Letters, IEEE 9(11):470–472

[18] Fear EC, Li X, Hagness SC, Stuchly M, et al (2002b) Confocal microwave imaging for breast cancer detection: Localization of tumors in three dimensions. Biomedical Engineering, IEEE Transactions on 49(8):812–822

[19] Guo L, Abbosh A (2015) Optimization-Based Confocal Microwave Imaging in Medical Applications. IEEE Transactions on Antennas and Propagation 63(8):3531–3539, DOI 10.1109/TAP.2015.2434394

[20] Hagness SC, Taflove A, Bridges JE (1998) Two-dimensional FDTD analysis of a pulsed microwave confocal system for breast cancer detection: Fixed-focus and antenna-array sensors. Biomedical Engineering, IEEE Transactions on 45(12):1470–1479

[21] Hagness SC, Taflove A, Bridges JE (1999) Three-dimensional FDTD analysis of a pulsed microwave confocal system for breast cancer detection: Design of an antenna-array element. Antennas and Propagation, IEEE Transactions on 47(5):783–791

[22] Hamaneh M, Chitravas N, Kaiboriboon K, Lhatoo S, Loparo K (2014) Automated Removal of EKG Artifact From EEG Data Using Independent Component Analysis and Continuous Wavelet Transformation. IEEE Transactions on Biomedical Engineering 61(6):1634–1641, DOI 10.1109/TBME.2013.2295173

[23] Haykin SS, Steinhardt AO (1992) Adaptive radar detection and estimation, vol 11. Wiley-Interscience

[24] Hyvarinen A (1999) Fast and robust fixed-point algorithms for independent component analysis. IEEE Transactions on Neural Networks 10(3):626–634, DOI 10.1109/72.761722

[25] Hyvärinen A, Oja E (2000) Independent component analysis: algorithms and applications. Neural Networks 13(4–5):411–430, DOI 10.1016/S0893-6080(00)00026-5

[26] Klemm M, Craddock I, Leendertz J, Preece A, Benjamin R (2008) Improved delay-and-sum beamforming algorithm for breast cancer detection. International Journal of Antennas and Propagation 2008

[27] Klemm M, Craddock I, Leendertz J, Preece A, Benjamin R (2009a) Radar-Based Breast Cancer Detection Using a Hemispherical Antenna Array ;Experimental Results. IEEE Transactions on Antennas and Propagation 57(6):1692–1704, DOI 10.1109/TAP.2009.2019856

[28] Klemm M, Leendertz J, Gibbins D, Craddock I, Preece A, Benjamin R (2009b) Microwave radar-based breast cancer detection: Imaging in inhomogeneous breast phantoms. Antennas and Wireless Propagation Letters, IEEE 8:1349–1352

[29] Klemm M, Craddock I, Leendertz J, Preece A, Gibbins D, Shere M, Benjamin R (2010) Clinical trials of a UWB imaging radar for breast cancer. In: 2010 Proceedings of the Fourth European Conference on Antennas and Propagation (EuCAP), pp 1–4

[30] Lazebnik M, McCartney L, Popovic D, Watkins CB, Lindstrom MJ, Harter J, Sewall S, Magliocco A, Booske JH, Okoniewski M, Hagness SC (2007) A large-scale study of the ultrawideband microwave dielectric properties of normal breast tissue obtained from reduction surgeries. Physics in Medicine and Biology 52:2637–2656

[31] Li J, Stoica P, Wang Z (2003) On robust Capon beamforming and diagonal loading. IEEE Transactions on Signal Processing 51(7):1702–1715, DOI 10.1109/TSP.2003.812831

[32] Li X, Pahlavan K (2004) Super-resolution TOA estimation with diversity for indoor geolocation. Wireless Communications, IEEE Transactions on 3(1):224 – 234, DOI 10.1109/TWC.2003.819035

[33] Li X, Davis S, Hagness S, Van Der Weide D, Van Veen B (2004) Microwave imaging via space-time beamforming: experimental investigation of tumor detection in multilayer breast phantoms. IEEE Transactions on Microwave Theory and Techniques 52(8):1856–1865, DOI 10.1109/TMTT.2004.832686

[34] Li X, Bond EJ, Van Veen BD, Hagness SC (2005) An overview of ultra-wideband microwave imaging via space-time beamforming for early-stage breast-cancer detection. Antennas and Propagation Magazine, IEEE 47(1):19–34

[35] Lim HB, Nhung NTT, Li EP, Thang ND (2008) Confocal microwave imaging for breast cancer detection: Delay-multiply-and-sum image reconstruction algorithm. Biomedical Engineering, IEEE Transactions on 55(6):1697–1704

[36] Lorenz R, Boyd S (2005) Robust minimum variance beamforming. Signal Processing, IEEE Transactions on 53(5):1684–1696

[37] Maklad B, Curtis C (2012) Neighborhood-based algorithm to facilitate the reduction of skin reflections in radar-based microwave imaging. Progress In Electromagnetics Research 39:115–139

[38] Martin J, Crowley JL (1995) Comparison of correlation techniques. In: International Conference on Intelligent Autonomous Systems, Karlsruhe (Germany), pp 86–93

[39] Maskooki A, Gunawan E (2009) Frequency domain skin artifact removal method for ultra-wideband breast cancer detection. Progress In Electromagnetics Research 98:299–314

[40] Nilavalan R, Gbedemah A, Craddock I, Li X, Hagness SC (2003a) Numerical investigation of breast tumour detection using multi-static radar. Electronics Letters 39(25):1787–1789

[41] Nilavalan R, Hagness SC, Veen BDV (2003b) Numerical Investigation of breast tumour detection using multi-static radar. IEEE Electronic Letters 39(25):1787–1789

[42] O'Halloran M, Glavin M, Jones E (2010a) Performance and robustness of a multistatic mist beamforming algorithm for breast cancer detection. Progress In Electromagnetics Research 105:403–424

[43] O'Halloran M, Jones E, Glavin M (2010b) Quasi-multistatic mist beamforming for the early detection of breast cancer. Biomedical Engineering, IEEE Transactions on 57(4):830–840

[44] O'Halloran M, Jones E, Glavin M (2010c) Quasi-Multistatic MIST Beamforming for the Early Detection of Breast Cancer. IEEE Transactions on Biomedical Engineering 57(4):830–840, DOI 10.1109/TBME.2009.2016392

[45] Piou J (2005) A state identification method for 1-d measurements with gaps. In: Proc. American Institute of Aeronautics and Astronautics Guidance Navigation and Control Conf

[46] Pope TL, Read ME, Medsker T, Buschi AJ, Brenbridge AN (1984) Breast skin thickness: normal range and causes of thickening shown on film-screen mammography. Journal of the Canadian Association of Radiologists 35(4):365–368

[47] Porter E, Kirshin E, Santorelli A, Coates M, Popovic M (2013) Time-domain multistatic radar system for microwave breast screening. IEEE Antennas and Wireless Propagation Letters 12:229–232, DOI 10.1109/LAWP.2013.2247374

[48] Sarafianou M, Craddock I, Henriksson T, Klemm M, Gibbins D, Preece A, Leendertz J, Benjamin R (2013) MUSIC processing for permittivity estimation in a Delay-and-Sum imaging system. In: 2013 7th European Conference on Antennas and Propagation (EuCAP), pp 839–842

[49] Sill JM, Fear EC (2005) Tissue sensing adaptive radar for breast cancer detection - experimental investigation of simple tumor models. Microwave Theory and Techniques, IEEE Transactions on 53(11):3312–3319, DOI 10.1109/TMTT.2005.857330

[50] Sutradhar A, Miller MJ (2013) In vivo measurement of breast skin elasticity and breast skin thickness. Skin Research and Technology 19(1):e191–e199, DOI 10.1111/j.1600-0846.2012.00627.x

[51] Van Veen BD, Buckley KM (1988) Beamforming: A versatile approach to spatial filtering. IEEE assp magazine 5(2):4–24

[52] Vorobyov SA (2013) Principles of minimum variance robust adaptive beamforming design. Signal Processing 93(12):3264–3277, DOI 10.1016/j.sigpro.2012.10.021

[53] Wang Y, Abbosh A, Henin B, Nguyen P (2014) Synthetic Bandwidth Radar for Ultra-Wideband Microwave Imaging Systems. IEEE Transactions on Antennas and Propagation 62(2):698–705, DOI 10.1109/TAP.2013.2289355

[54] Winters DW, Bond EJ, Hagness SC (2006) Estimation of the frequency-dependent average dielectric properties of breast tissue using a time-domain inverse scattering technique. IEEE Transactions on Antennas and Propagation 55(11):3517–3528

[55] Woody CD (1967) Characterization of an adaptive filter for the analysis of variable latency neuroelectric signals. Medical and biological engineering 5(6):539–554, DOI 10.1007/BF02474247

[56] Xie Y, Guo B, Xu L, Li J, Stoica P (2006) Multistatic adaptive microwave imaging for early breast cancer detection. Biomedical Engineering, IEEE Transactions on 53(8):1647

[57] Yin T, Ali F, Reyes-Aldasoro C (2015) A Robust and Artifact Resistant Algorithm of Ultra-wideband Imaging System for Breast Cancer Detection. IEEE Transactions on Biomedical Engineering 62(6):1514–1525, DOI 10.1109/TBME.2015.2393256

[58] Zhi W, Chin F (2006) Entropy-based time window for artifact removal in UWB imaging of breast cancer detection. Signal Processing Letters, IEEE 13(10):585–588

Chapter 5
Tumour Classification

Raquel Cruz Conceição, Marggie Jones, Panagiotis Kosmas, and Yifan Chen

Recent breast tissue dielectric spectroscopy measurements in [35] suggest that the malignant-to-benign dielectric contrast may not be sufficiently high to allow for tumour classification based on backscatter intensity. Alternatively, it is well known that the architectural distortion in breast parenchyma can aid in distinguishing malignant tumours from benign masses [56, 65]. Mammographic image analysis shows that malignant tumours usually have an irregular shape and are surrounded by a radiating pattern of linear spicules, also their margins are obscured and indistinct. Conversely, benign tumours are roughly elliptical and usually have well-circumscribed margins [56, 65]. Accordingly, microwave *backscatter signature*— the signal which is reflected when a target is illuminated by microwaves—could be potentially useful for discrimination between benign and malignant tumours, and for inferring their size. Tomographic image reconstruction methods, which solve the inverse scattering problem to obtain a coarse estimation of the breast dielectric profile, have been used in this context [19].

R.C. Conceição (✉)
Faculdade de Ciências, Instituto de Biofísica e Engenharia Biomédica, Universidade de Lisboa, Lisbon, Portugal
e-mail: raquelcruzconceicao@gmail.com

M. Jones
Medical Device Research Group, Translational Research Facility, National University of Ireland, Galway, Ireland
e-mail: marggie.jones@nuigalway.ie

P. Kosmas
King's College London, Strand, London, UK
e-mail: panagiotis.kosmas@kcl.ac.uk

Y. Chen
South University of Science and Technology of China, Shenzhen, China
e-mail: chen.yf@sustc.edu.cn

© Springer International Publishing Switzerland 2016
R.C. Conceição et al. (eds.), *An Introduction to Microwave Imaging for Breast Cancer Detection*, Biological and Medical Physics, Biomedical Engineering, DOI 10.1007/978-3-319-27866-7_5

In this chapter, the following will be reviewed: numerical morphological models to represent breast tumours, some of the recent UWB tumour classification techniques using the early-time response [73], the late-time response [8–10], the decomposition of the time reversal operator (DORT, from the French) [33] and using pattern recognition algorithms to extract the radar target signature of tumour in microwave backscatter before they are classified using linear discriminant analysis, quadratic discriminant analysis, support vector machines, spiking neural networks and self-organising maps [12–14, 17, 44, 51].

The outline of this chapter is as follows: firstly, the tumour numerical models that have been used in the literature are presented; then the classification of early-stage breast cancer using early-time response, late-time response, classification on contrast-enhanced tumours and classification based on radar target signature of tumours are addressed; finally the classification of early-stage breast tumours in experimental setups is presented.

5.1 Tumour Numerical Morphological Models

Segmentation of tumours in medical images is a difficult task. Several approaches to modelling the tissue anomalies have appeared in the literature. For example, spherical harmonic functions have been applied to model the irregular shape of tumours [19]. Two other mathematical approaches which have been used in the literature to model breast tumours will be presented in the following Sects. 5.1.1 and 5.1.2.

5.1.1 Polygonal-Based Models

In [56], Rangayyan et al. have chosen to approximate the mammographic tumour boundaries by polygons. It was observed that the polygonal approximations match the hand-drawn diseased tissue boundaries input by the radiologist very well. The approach considered in [8–10, 33, 73] follows [56], where the tumour boundaries are approximated by polygons. The first stage is to establish the elliptical behaviour of tissue anomalies. The baseline ellipse is defined in polar coordinates as

$$B(\varphi) = \frac{ab}{\sqrt{a^2 \sin^2 \varphi + b^2 \cos^2 \varphi}} \tag{5.1}$$

where a and b are the semi-major and semi-minor axes, and φ is the angle as indicated in Fig. 5.1. The next step is to modify the initial shape to produce the proper mass border with various levels of irregularity. The simulation routine consists of the following stages:

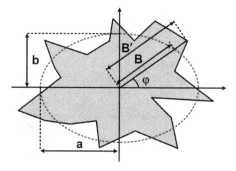

Fig. 5.1 Border deviation profiles indicating how to generate a tumour border from a baseline ellipse, in [10]

1. Define the number of sides of the polygonal approximation to the tumour boundary, Q.
2. Define the distribution function of φ. For simplicity,

$$\varphi_q \in \mathscr{U}[0, 2\pi], \quad (q = 1, 2, \cdots, Q), \tag{5.2}$$

where $\mathscr{U}$ denotes a uniform distribution.
3. For each φ_q, define the distribution of the border deviation profile $\zeta(\varphi_q)$ and apply it to the elliptical profile in a multiplicative fashion as (see also Fig. 5.1):

$$B'(\varphi_q) = B(\varphi_q)\left[1 + \zeta(\varphi_q)\right]. \tag{5.3}$$

It is assumed that $\zeta \in \mathscr{U}[-\Delta B, +\Delta B]$ and ζ is independent of φ_q. The two parameters, Q and ΔB, determine the ruggedness of the tumour boundary. As $Q \to +\infty$ and $\Delta B \to 0$, the mass border approaches a perfect ellipse. If a binary classification of the tumour morphology, as in [9, 10, 73], is assumed, the shape of each random target is assumed to fall into one of the following two categories: Oval/MAcrolobulated (OMA) or MIcrolobulated/SPiculated (MISP), which are related to the shape and margin descriptors used in mammography [17]. The OMA class includes targets that correspond to round, oval or macrolobulated shape descriptors. Inversely, the MISP class exhibits fine-scale undulations over the target surface or spicules radiating from the body of the target. OMA and MISP shapes often correspond to benign and malignant tumours, respectively. As observed in [8], the circumference texture becomes considerably rugged as Q decreases and ΔB increases. Therefore, the ranges of Q and ΔB could be carefully chosen to reflect the distinctive features of these two types of anomalies.

Figure 5.2 plots the mass border deviations corresponding to the OMA and MISP tumours. Eventually, some intermediate stages are required to correlate the morphology classification outputs with the final breast cancer screening result.

A more recent study by Oliveira et al. [52, 53] has further extended this modelling technique to develop tumours in 3-D. Firstly, they assume that the shape of a tumour resembles an ellipsoid, which is modelled as a 3-D polygon consisting of a series of triangular faces interlinked together. The number of faces in the polygon defines the

Fig. 5.2 Examples of the
OMA and MISP boundary
profiles. The OMA target is
simulated by setting
$Q \sim \mathcal{U}[80, 100]$ and
$\Delta B \sim \mathcal{U}[0.1, 0.3]$, whereas
the MISP target is simulated
by setting $Q \sim \mathcal{U}[10, 30]$ and
$\Delta B \sim \mathcal{U}[0.8, 1]$, in [9, 10]

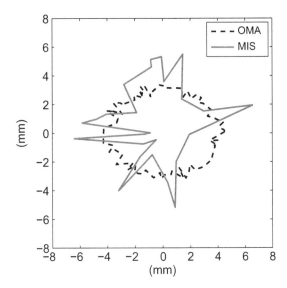

number of spicules the model will have.

$$\frac{d^2 \cos^2 \theta \sin^2 \varphi}{a^2} + \frac{d^2 \sin^2 \theta \cos^2 \varphi}{b^2} + \frac{d^2 \cos^2 \varphi}{c^2} = 1 \qquad (5.4)$$

where (d, θ, φ) are the spherical coordinates that describe the ellipsoid, d is the distance from the vertices of each triangular face to the centre of the polygon, and (a, b, c) are the lengths for each semi-axis. Similarly to [8], a distortion is applied to each vertex of the polygon by modifying the distance from the centre to each vertex

$$d'(\theta, \varphi) = a[d(\theta, \varphi)(1 + S(\theta, \varphi))] \qquad (5.5)$$

in which d is the new distance of each vertex in the triangular face to the centre of the ellipsoid. S varies between 0 and 1, which will make the ellipsoid vary between smooth (equivalent to OMA category) and spiculated (equivalent to MISP category), respectively. Additionally, the authors in [53] have created tumour models in which the whole surface or only part of the surface has been covered by spicules. Figure 5.3 shows polygon-based models with an even distribution of spiculation on the left (a, c and e) and an uneven distribution of spiculation on the right (b, d and f). The authors have further developed different levels of models with unevenly distribution of spicules: the spiculation is restricted to one part of the surface of the tumour

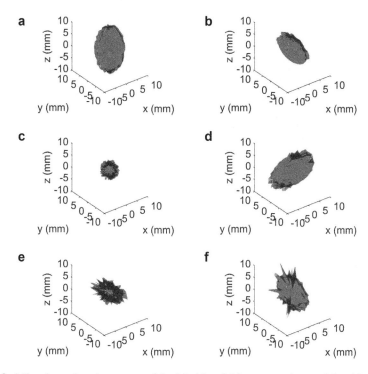

Fig. 5.3 3-D polygon-based tumour models. (**a**), (**c**) and (**e**) correspond to models with an even distribution of spiculation while (**b**), (**d**) and (**f**) correspond to models with an uneven distribution of spiculation. Courtesy of Bárbara Oliveira, Martin O'Halloran, Raquel Conceição, Martin Glavin and Edward Jones, after [52]

model (Fig. 5.3b), concentration of spicules in non-adjacent parts of the surface of the tumour model (in Fig. 5.3d) and varying levels of spiculation at different parts of the surface of the tumour model (in Fig. 5.3f).

5.1.2 Gaussian Random Spheres

Gaussian random spheres (GRSs) have been used to generate benign and malignant tumour models, following an algorithm proposed by Muinonen et al. [47, 48] originally designed for modelling asteroids and comets in an astrophysics context. GRSs can be easily modified to provide different sizes, shapes and textures of surface in 3-D, which are the characteristics that most significantly influence the RTS of tumours which are used for classification purposes [17]. The shape of the GRS is given by a radius vector, $r = r(\vartheta, \varphi)$, which is defined by the logradius, or logarithmic radius, $s = s(\vartheta, \varphi)$

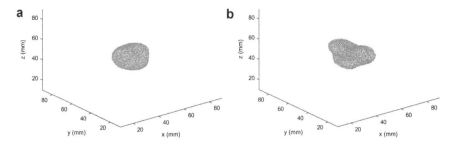

Fig. 5.4 Samples for different Gaussian random spheres (GRSs) representing benign tumour models with an average radius size of 5 mm: (**a**) smooth and (**b**) microlobulated. Images courtesy of Raquel Conceição, Martin O'Halloran, Martin Glavin and Edward Jones, The Electromagnetics Academy and Taylor & Francis Ltd. (www.tandfonline.com), after [11–13, 15]

$$r(\vartheta, \varphi) = \alpha \cdot \exp\left[s(\vartheta, \varphi) - \frac{1}{2}\beta^2\right] \tag{5.6}$$

$$s(\vartheta, \varphi) = \sum_{l=0}^{\infty} \sum_{m=-l}^{l} s_{lm} Y_{lm}(\vartheta, \varphi). \tag{5.7}$$

5.6 and 5.7, (ϑ, φ) stand for the spherical coordinates, α is the mean radius, β is the standard deviation of the logradius, Y_{lm} are the orthonormal spherical harmonics, s_{lm} are the spherical harmonics weight coefficients, in which l and m stand for the degree and the order of expansion, respectively [47].

In previous studies [12–14, 17, 44, 51], a total of four different shapes and four different sizes of tumour models have been considered. Malignant tumours are represented by spiculated and microlobulated GRSs (equivalent to MISP category), benign tumours are divided into two type categories: smooth and microlobulated (equivalent to OMA category). The average radii of all shapes of spheres are 2.5, 5, 7.5 or 10 mm. Microlobulated, macrolobulated and smooth GRSs are obtained by varying the correlation angle: between 5° and 20° for microlobulated GRSs, between 25° and 45° for macrolobulated GRSs and between 50° and 90° for smooth GRSs. Spiculated GRSs are obtained by adding 3, 5 or 10 spicules to smooth GRSs, as previously used in [17]. Examples of benign tumour models based on the GRSs method, with radii of 5 mm, are shown in Fig. 5.4. Examples of malignant tumour models (including microlobulated and spiculated tumour models with: 3, 5 and 10 spicules), created with the GRS method, with radii of 5 mm, are shown in Fig. 5.5.

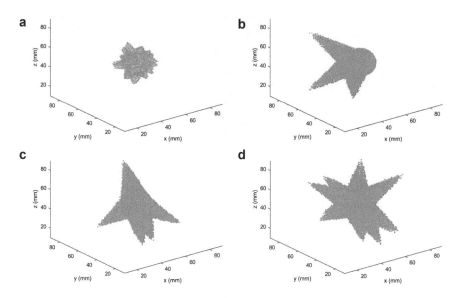

Fig. 5.5 Samples for different Gaussian random spheres (GRSs) with an average radius size of 5 mm, representing malignant tumour models: (**a**) microlobulated, (**b**) spiculated with 3 spicules, (**c**) spiculated with 5 spicules and (**d**) spiculated with 10 spicules. Images courtesy of Raquel Conceição, Martin O'Halloran, Martin Glavin and Edward Jones, and The Electromagnetics Academy, after [11, 15]

5.2 Classification of Early-Stage Breast Cancer in Numerical Simulations

In this chapter, the different classification approaches used to classify numerically simulated tumours, detailed in Sect. 5.1, are presented. These include classification based on early-time response, late-time response, contrast-enhanced classification and classification based on RTS extraction.

5.2.1 Tumour Classification Based on Early-Time Response

It has been shown, in [73], that the early-time UWB backscatter signal is affected by the morphology of the scatterer. Different scattering points on the surface of the scatterer will alter the shape of the early-time response. A rough tumour with multiple spicules has more scattering points than a benign tumour with a compact shape [17].

The backscatter signals are calculated from a two-dimensional (2-D) Finite-Difference Time-Domain (FDTD) TE_z mode simulation. A simplified cylindrical breast model is used, as shown in Fig. 5.6. The clutter model used here follows that

Fig. 5.6 Setup of the 2-D
FDTD lattice with the centre
sensor acting as a transceiver.
The rest of the receiving
sensors are placed at equal
distance from the tumour at
regular intervals, in [73]

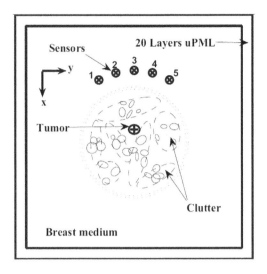

described in [8], where tissue inhomogeneity can be identified to occur in regions, and a number of different sized clutter sources are randomly placed in each of these regions. For the FDTD calculation, 0.5 mm square grids are used to discretise the computational space. The Courant factor, C, is chosen to be 0.5 for numerical stability. The computational space is a square box, with 401 cells along the direction of propagation and along the direction of electric field source. A Uniaxial Perfectly Matched Layer (UPML) is used beyond the space to minimise reflections from the boundaries of the simulation box. The single-pole Debye model is used to simulate the breast medium, clutter and target tumour. The Debye medium is incorporated into the FDTD equations through the time-domain auxiliary differential equations. A point source (Sensor 3 in Fig. 5.6) is placed such that the centre of the breast is 6.5 cm away from the source. This transmitter also functions as a receiver. Two additional point receivers are placed to the left (Sensors 1 and 2 in Fig. 5.6) and to the right (Sensors 4 and 5 in Fig. 5.6) of the transmitter in 15° intervals with reference from the centre line passing through Sensor 3 and the centre of the breast. The excitation signal applied to the transmitter is a modulated Gaussian pulse.

The concept of a correlator at the receiver end is adopted. It relies on a template signal for correlation with the backscatter response. The phenomenon of interest here is the amount of deformation of the specular returns from different types of tumours. Correlation can then be used as a mean of quantifying the differences. The template signal can be derived by passing the signal through a phantom material with no scatterer present. Consider the backscatter response from a tumour target. Let $E(t)$ be the *clean* signal impinging on a receiver from a transmitter. The backscatter signal $E_{\text{back}}(t)$ can be modelled as

$$E_{\text{back}}(t) - \pm \alpha_1 E(t - \tau_1) \pm \alpha_2 E(t - \tau_2) + E_{\text{late}}(t) \qquad (5.8)$$

Fig. 5.7 Sample received signals at Sensor 3 for a tumour target. Three portions can be identified. The 1st reflection has undergone an inversion, in [73]

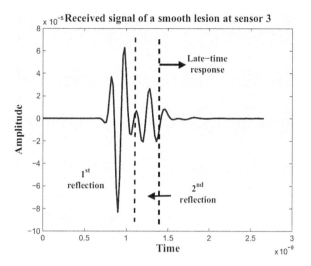

where $\alpha_{1,2}$ are the attenuation factors, and $\tau_{1,2}$ are the delays due to reflections that occur during the entry and exit of the wave from the tumour, respectively, and $E_{\text{late}}(t)$ is the late-time response of the tumour. The $\pm$ signs denote possibility of inversion of the signal due to the reflection coefficient between the two mediums.

This can be observed in Fig. 5.7, where 3 different regions can be observed distinctly with a small amount of overlapping regions. Subsequently, the correlation can be computed following the procedures in [73]. It is expected that as the surface of a tumour gets rougher, there will be a higher chance that the early response of the received signal from any of the sensors is different. The degree of difference increases as the roughness increases. The tumours are consequently classified according to the following criteria [73].

Let $C_i, i \in \{1, 2, \cdots, P\}$, denote the P peak correlation coefficients obtained from the array of sensors from a tumour target. The 4 measures include:

1. Mean of the peak correlations, $M_1 = \sum_{i=1}^{P} C_i / P$.
2. Range of the peak correlations, $M_2 = \max(C_i) - \min(C_i)$.
3. Standard deviation of the peak correlations, $M_3 = \sqrt{\sum_{i=1}^{P} (C_i - M_1)^2 / (P - 1)}$.
4. Maximum deviation of the peak correlations with respect to the mean value, $M_4 = \max(|C_i - M_1|)$.

A total of 60 tumours are simulated using the tumour generator described in 5.1.1. The tumour database comprises 2 groups (OMA and MISP) of 30 tumours each. The backscatter signal at the 5 different locations is recorded as shown in Fig. 5.6. The early-time responses are processed and the 4 criteria for tumour classification are evaluated. In Fig. 5.8, the mean value (M_1) is used as a tissue discrimination indicator. It is observed that overall the data points associated with the smooth and rough tumours are well separated. Similar trends have been observed when other measures are evaluated. Hence, the correlator classifiers proposed in [73] exhibit the potential in differentiation between tumours with smooth and spiculated boundaries.

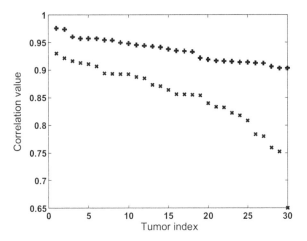

Fig. 5.8 Tumour
classification based on the 1st
measure. '*plus*' denote OMA
tumours and '*cross*' denote
MISP tumours, in [73]

5.2.2 Tumour Classification Based on Late-Time Response

The following subsections address tumour classification based on late-time response
in a monostatic and a multistatic radar systems simulations.

5.2.2.1 Complex Natural Resonance of Late-Time Response

In [8], a monostatic radar configuration is considered, where the observation point
coincides with the source point. The incident field is a plane wave E^{inc} parallel to the
cylindrical tumour and the z axis as illustrated in Fig. 5.9. The surrounding region is
the baseline breast medium with a wavenumber k_b. On the other hand, the tumour
has a wavenumber k_l. A numerical approach, following [58], is applied to derive the
backscatter pulse template. The dielectric cylinder is divided into a number of square
cells, which are small enough so that the electric field intensity is nearly uniform in
each cell, as shown in Fig. 5.9. A system of linear equations is then obtained by
enforcing, at the centre of each cell, the condition that the total field must equal
the sum of the incident and scattered fields. By solving this set of equations, the
scattered field at the observation point can be obtained. It is further assumed in [8]
that the sensor excitation is a differentiated Gaussian monocycle

$$E^{inc}(t) = \frac{t}{T} \exp\left(-\frac{t^2}{4T^2}\right) \quad (V/m) \tag{5.9}$$

where T controls the width of the excitation.

The normalised transient response due to the presence of a high-water-content
breast mass in a fatty tissue environment is shown in Fig. 5.10. As can be observed
in Fig. 5.10, the backscatter pulses are considerably different for different values
of Q and ΔB. In general, the impulse response from a target can be expressed

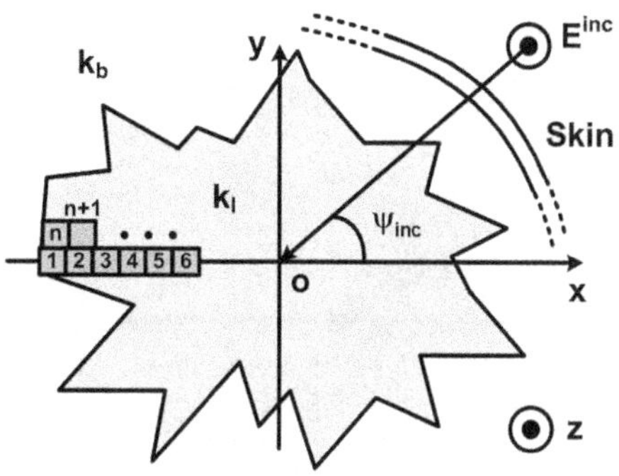

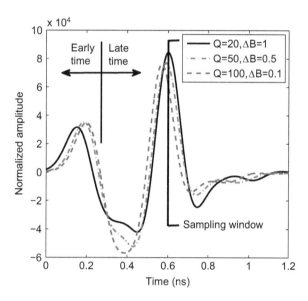

Fig. 5.9 A cross-sectional view of the plane wave impinging on a dielectric cylinder with a stipulated border profile, in [8]

Fig. 5.10 Time-domain characteristics of the backscatter signals for tumours with various levels of morphological irregularity. Also indicated are the early-time and late-time responses as well as the sampling window for CNR extraction, in [8]

as the sum of the early-time response, which contains only singularities of the excitation source, and the late-time scattering characteristics, which start to build up after the target is illuminated by a transient incident field and stabilises after the multiplied reflected fields form the damped oscillations. The late-time response can be expanded using the total least square Prony's method [7, 8]. Each non-zero singularity corresponds to a complex natural resonance (CNR), which is related to the target geometric signature. The early-time and late-time responses are indicated in Fig. 5.10. It can be observed that the late-time response due to the natural oscillations of the scatterer is dependent on Q and ΔB, thus containing useful information about the morphological features. Finally, the sampling window for CNR extraction is illustrated in Fig. 5.10. The damping factors for tumours with various levels of shape irregularity are shown in Fig. 5.11. As can be observed in this figure, the resonant frequencies corresponding to different Q and ΔB are very similar. One dominant pole is located between 1 and 2 GHz and the other is

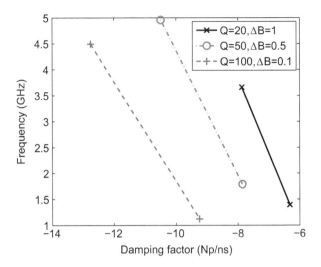

Fig. 5.11 Damping factors for tumours with various levels of morphological Irregularity, in [8]

located between 3.5 and 5 GHz. On the one hand, for tumours with the same level of spiculatedness, the high-frequency pole exhibits a smaller attenuation factor as compared to the low-frequency one. On the other hand, the damping factors for both the two poles increase as the tumour boundary becomes more deviated from the baseline ellipse (i.e. with lower Q and higher ΔB). In general, the poles are well separated in the s-plane and therefore classification of tumour morphologies seems plausible.

5.2.2.2 Multiple-Input Multiple-Output Radar for Tumour Classification

A monostatic radar architecture does not allow for the full exploitation of waveform and spatial diversity gains, which can be achieved with a Multiple-Input Multiple-Output (MIMO) radar [25, 36] and may be useful for enhanced tissue discrimination. The study in [10] extends the methodology in [8] by employing a MIMO radar for tumour classification, where each antenna takes turns to transmit a UWB pulse and all the antennas are used to record the backscatter. Also, different excitations are used at different transmit antennas to probe the breast. The corresponding system setup is illustrated in Fig. 5.12a. The incident waveform transmitted from the kth array element, E_k^{inc} ($k = 1, 2, \cdots, K$), is chosen from the family of Modulated and Modified Hermite Polynomials (MMHPs) [23]

$$E_k^{\text{inc}} = h_{\Omega_k}\left(\frac{t}{T}\right) \exp\left(-\frac{t^2}{4T^2}\right) \cos\left(2\pi f_k t\right) \quad (\text{V/m}) \tag{5.10}$$

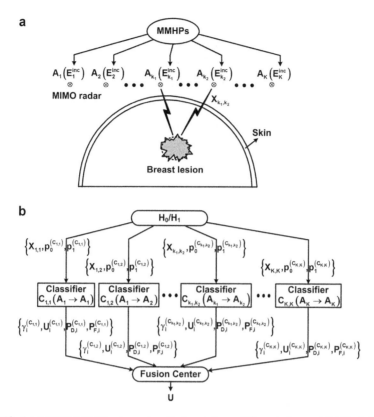

Fig. 5.12 (a) A UWB MIMO radar system for breast tissue differentiation, and (b) block diagram of a data-fusion-based tumour classification system, where multiple classifiers correspond to various spatial diversity paths offered by the MIMO radar, in [10]

where $\Omega_k = 0, 1, 2, \cdots$, h_{Ω_k} is the Ω_k^{th}-order Hermite polynomial, T controls the duration of the pulses and f_k is the carrier frequency. The multiple transmitted MMHP waveforms are orthogonal to each other and have nearly constant pulse duration and bandwidth regardless of the pulse order. The relevant phenomenon-of-interest observed at Antenna A_{k_2} induced by the probing signal from Antenna A_{k_1} is the damping factor of a specific CNR, X_{k_1,k_2}. As shown in [8], X_{k_1,k_2} is a random variable due to the non-deterministic nature of the tumour shape. Furthermore, the statistical properties of X_{k_1,k_2} vary with the border profiles of tumours, which in turn are correlated with the pathologic features of the tissue. The canonical tumour classification problem can thus be stated as follows (see also Fig. 5.12b for a block diagram of the MIMO radar system architecture).

Let there be K antennas $A_1, \cdots, A_K$ in the MIMO radar imaging system as depicted in Fig. 5.12a. Antenna A_{k_2} observes a random damping factor X_{k_1,k_2} derived from the backscatter due to the excitation waveform transmitted from Antenna A_{k_1} ($1 \leq k_1, k_2 \leq K$). A virtual classifier C_{k_1,k_2} corresponding to the diversity path

$A_{k_1} \rightarrow A_{k_2}$ is defined. Each X_{k_1,k_2} is then transformed to a decision message by C_{k_1,k_2} as $U_i^{(C_{k_1,k_2})} = \gamma_i^{(C_{k_1,k_2})}(X_{k_1,k_2})$ using the th local mapping $\gamma_i^{(C_{k_1,k_2})}$

$$U_i^{(C_{k_1,k_2})} = \begin{cases} -1, & \text{if } H_0 \text{ is accepted} \\ 1, & \text{if } H_1 \text{ is accepted} \end{cases} \tag{5.11}$$

where H_0 and H_1 represent the two respective hypotheses: 'Target is OMA' and 'Target is MISP'. The Probability Density Functions (PDFs) of X_{k_1,k_2} are $p_0^{(C_{k_1,k_2})}$ and $p_1^{(C_{k_1,k_2})}$ under H_0 and H_1, respectively. If $p_0^{(C_{k_1,k_2})}$ or $p_1^{(C_{k_1,k_2})}$ cannot be specified completely (e.g. due to limited empirical data for reliable statistical modelling of the damping factor), the PDF should be specified as belonging to an uncertainty class, $p_0^{(C_{k_1,k_2})} \in \mathscr{P}_0^{(C_{k_1,k_2})}$ or $p_1^{(C_{k_1,k_2})} \in \mathscr{P}_1^{(C_{k_1,k_2})}$. The local decisions $U_i^{(C_{k_1,k_2})}$, together with their corresponding probability of detection $P_{D,i}^{(C_{k_1,k_2})}$ and probability of false alarm (or false positives) $P_{F,i}^{(C_{k_1,k_2})}$ subject to each mapping rule $\gamma_i^{(C_{k_1,k_2})}$, are sent to a data fusion centre. The fusion centre then makes a global decision U ($U = -1$ for accepting H_0 and $U = 1$ for accepting H_1). The following two issues need to be addressed: (1) the estimation of statistical properties of X_{k_1,k_2} under H_0 or H_1, and (2) the optimal rules at the fusion centre to achieve robust tumour classification.

Regarding (1), the statistical properties of X_{k_1,k_2} under H_0 or H_1, the empirical Cumulative Distribution Functions (CDFs) of the damping factors are derived through comprehensive numerical simulations [10]. Some typical results are illustrated in Fig. 5.13. As shown in this figure, the damping factors for OMA tumours fall within a narrow range. Specifically, the Laplace distribution yields the best fit to the empirical data. Subsequently, a truncated Laplace distribution p_0 is fit to the simulated data. Contrary to the OMA class, the damping factors for MISP tumours are scattered over a much wider range due to more ill-defined target geometries. The empirical results could be fit with a general Laplace PDF p_1. It is more difficult to characterise p_1 exactly due to the limited simulated data and the widely spread damping factors. Hence, it is supposed that p_0 is completely specified, while p_1 belongs to an uncertainty class $\mathscr{P}_1$, which is defined through the CDF bounds plotted in Fig. 5.13. Note that the CDF curves for OMA and MISP data are well separated, which demonstrate the feasibility of tumour discrimination through statistical inference of the damping factors.

Regarding (2), the optimal rules at the fusion centre to achieve robust tumour classification, the asymptotic regime will be considered, where the number of randomly generated tumours goes to infinity. In this case, a suitable way to compare fusion schemes is through the error exponent measure. The best achievable asymptotic exponent for the probability of error is given by the Chernoff information under the Bayesian framework. The goal is thus to design optimal fusion rules that maximise this exponent. Subsequently, two fusion schemes are proposed, namely the selection combining fusion and log likelihood ratio based fusion, for discrimination between benign and malignant masses, in [10]. One conventional

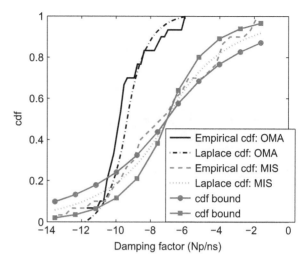

Fig. 5.13 Cumulative distribution functions of OMA and MISP data, in [10]

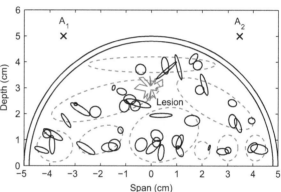

Fig. 5.14 A numerical breast phantom with clustered scatterers. Each cluster, encircled by an ellipse, contains a number of clutter items. The centres of clutter sources are uniformly distributed in each elliptical area. Different dielectric variations have been assigned to different clusters as detailed in [8], in [10]

way to design decision rules when the CDF of the observations is not specified completely is the *minimax* approach [75], where the goal is to optimise the worst-case performance over the uncertainty class. However, this approach will be of little practical significance if it is of low likelihood. In [10], a more statistically relevant solution is proposed to optimise the classifier performance along the receiver outage error exponent characteristics curve, where the error exponent is defined for a sufficiently low outage probability.

A simplistic 2×2 MIMO radar architecture as shown in Fig. 5.14 is considered. Antenna A_1 transmits an MMHP pulse with $\Omega_1 = 1, f_1 = 1$ GHz and $T = 50$ ps, while Antenna A_2 transmits a pulse with $\Omega_2 = 1, f_2 = 4$ GHz and $T = 50$ ps. A numerical breast phantom with clustered clutter distributions is employed for simulation studies [8–10]. Figure 5.15 depicts the Receiver Operating Characteristic (ROC) curves when the outage probability is 10 %. It can be seen that no single diversity path produces superior classification performance over the whole range of probability of detection. Figure 5.15 also shows the maximum realisable ROC

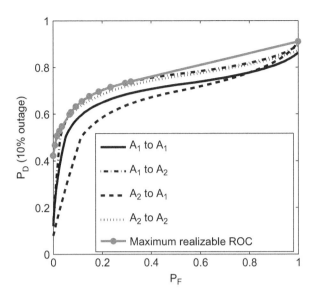

Fig. 5.15 ROC curves at 10 % outage probability, in [10]

predicted by the convex hull containing other ROC curves [66], which yield the best true positive rates over the entire ROC space. The vertex points on the hull correspond to the existing classifiers. The results in Fig. 5.15 clearly demonstrate the advantage of a MIMO radar for tumour classification, which fully explores the information of the phenomenon of interest provided by different classifiers (through waveform diversity and spatial diversity) to improve the overall classification performance.

5.2.3 Contrast-Enhanced Tumour Classification

In this section, initial results of classification when contrast is used to enhance the tumour response are shown.

5.2.3.1 Contrast-Enhanced Tumour Classification

The tissue differentiation capability will deteriorate severely if the intrinsic contrast between the dielectric properties of tumour and normal tissues is small. This phenomenon has been reported in [35]. According to this study, the contrast in the dielectric properties between tumour and normal glandular/fibroconnective tissues (non-adipose) is no more than about 10 %. As nearly all breast tumours originate within glandular tissues, the tumour, whether malignant or benign, will be a weakly scattering target within a highly cluttered environment. This would lead to severe deterioration of tumour classification performance [8].

The problem can be possibly overcome by the infusion of contrast agents such as dielectric or conducting micro/nanoparticles into a tumour site (see, e.g., [42, 43, 69]), which can provide comparative information for tissue discrimination through the monitoring of backscatter signature over time. Recent preliminary studies have suggested that the dielectric properties of a tumour are altered in the presence of air-filled microbubbles or Single-Walled Carbon Nanotubes (SWCNTs) [42, 43]. In [9], the use of contrast agents for classifying a tumour as malignant or benign is studied, by analysing the CNRs of the differential backscatter responses before and after the deployment of contrast agents to the suspicious site.

It is assumed that the contrast agent will accumulate in the tumour without altering its border profile. To simulate the reduction or increase in tissue properties caused by randomly positioned microbubbles or SWCNTs, the polygonal tumour is firstly divided into a number of square cells that are small enough so that the electric field intensity is nearly uniform in each cell. Subsequently, the post-contrast relative permittivity and conductivity assigned to each cell are assumed to be $\varepsilon_{post} = \varepsilon_b (1 + \gamma_{pre}) (1 + \nu)$ and $\sigma_{post} = \sigma_b (1 + \gamma_{pre}) (1 + \nu)$ over the entire frequency spectrum, where ε_b and σ_b represent the baseline tissue properties. The parameter γ_{pre} determines the pre-contrast dielectric values of the tumour. The parameter ν is a random variable generated according to $\nu \sim \mathscr{U}[\nu_{min}, \nu_{max}]$. The experimental investigation shows that the range of ν is within 40 % for microbubble concentrations typically administered in human ultrasound procedures (i.e. within the 10^9 microbubbles/mL dosage) [42]. For SWCNTs, it is assumed that there is a similar range of ν following the dielectric spectroscopy data in [43] for various concentrations of SWCNTs in water. Figure 5.16 illustrates the post-contrast dielectric profiles of the tumour tissue. This method of modelling the tumour dielectric properties could be readily integrated into the numerical approach in [8] to derive the backscatter microwave signature.

To study the efficacy of microwave imaging with varying concentrations of contrast agents, the following simulation protocol is used. For the th target realisation belonging to either OMA or MISP category, the following are

Fig. 5.16 The heterogeneous post-contrast dielectric profile of the target for $\nu \sim \mathscr{U}[-35\%, -25\%]$. The background medium represents the baseline normal breast tissue, where the dielectric property is normalised to 1, in [9]

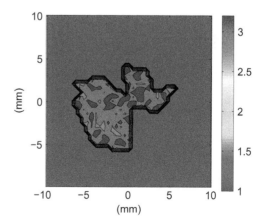

recorded: the pre-contrast breast response E_i^{pre} and the post-contrast responses with varying dielectric alterations $E_{i,j}^{post}$ ($j = 1, 2, 3, 4$), which correspond to $\nu_1 \sim \mathcal{U}[-15\,\%, -5\,\%]$, $\nu_2 \sim \mathcal{U}[-35\,\%, -25\,\%]$, $\nu_3 \sim \mathcal{U}[5\,\%, 15\,\%]$ and $\nu_4 \sim \mathcal{U}[25\,\%, 35\,\%]$, respectively. The first two ensembles of dielectric profiles may represent two volume fractions of microbubble inclusions, while the last two may represent two concentrations of SWCNTs. The CNRs of E_i^{pre} and the differential backscatter responses $\Delta E_{i,j} = E_{i,j}^{post} - E_i^{pre}$ are then extracted following the procedures in [8]. Similar to [8], two most dominant resonances can be obtained from the backscatter. The high-frequency resonance is applied for the analysis as it provides better classification performance as compared to the low-frequency component.

Typical results for the empirical CDFs of the damping factors in the case of microbubble inclusions are presented in Fig. 5.17 (post-contrast differential backscatter). A Laplace CDF is applied to fit the post-contrast data. It can be seen from Fig. 5.17 that, at the post-contrast stage, the damping factors for OMA tumours fall within a narrower range as compared to those for MISP targets. Furthermore, the CDF curves for OMA and MISP data are separated in the CDF space, which demonstrates the feasibility of tumour discrimination through statistical inference of the damping factors. On the contrary, at the pre-contrast stage, the damping factors are almost uniformly distributed irrespective of the tissue border profiles, thereby resulting in a non-discriminative classifier.

Using the above model, the tissue discrimination step is pursued. The formulated detection problem aims to distinguish between the null hypothesis H_0 ('Target is OMA') and the alternative hypothesis H_1 ('Target is MISP'). The Neyman–Pearson (NP) criterion combined with Youden's index is applied for discrimination

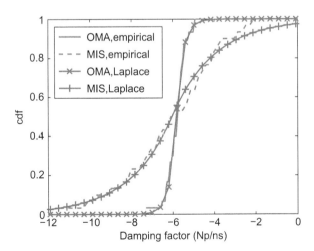

Fig. 5.17 Cumulative distribution functions of OMA and MISP data for microbubble inclusions causing dielectric reduction, in [9]

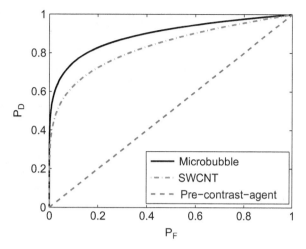

Fig. 5.18 ROC curves for the pre- and post-contrast-agent tumour classification, in [9]

purposes [9]. Figure 5.18 depicts the resulting ROC curves at both the pre- and post-contrast stages. It can be seen that the introduction of the contrast agents produces much better classification performance as compared to the pre-contrast stage. Indeed, the latter yields the most ineffective random-guess classifier with ROC lying along the diagonal line connecting $(0, 0)$ and $(1, 1)$.

5.2.3.2 Contrast-Enhanced Tumour Classification Based on the DORT Method

The DORT technique [55] has been shown to achieve selective focusing in the presence of multiple tumours using a 2-D MRI-based breast model under the assumption that the tumour response is fully available [32]. Tumour classification via the DORT method is based on the difference in the singular value spectrum of the Multistatic data Matrix (MDM), which is caused by tumours of different size and shape [33]. Since perfect knowledge of the tumour-backscattered signal is impossible in practice, the differential data obtained by simulations before and after the injection of contrast agents is considered, similarly to the previous section.

The DORT method is based on an eigenvalue decomposition of the Time Reversal Operator (TRO), or equivalently, a Singular Value Decomposition (SVD) of the MDM. The MDM is constructed by having each element of the antenna array transmit a signal and record the response of the system at all other antennas. A 2-D numerical MRI-based breast phantom with updated low-adipose tissue properties to reflect the low contrast (as low as 10 %) of malignant to normal fibroglandular tissue is used. The dielectric constant distribution of the breast model at 1 GHz is shown in Fig. 5.19. MISP and OMA tumours of size comparable to a 1, cm-diameter spherical tumour are introduced inside the fibroglandular region, where they most commonly occur. The contrast of these tumours to the surrounding tissue is of the order of 10 %. Simulated data is generated for TM_x polarisation using the FDTD

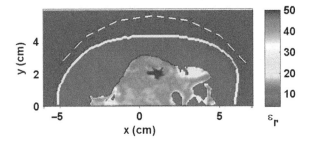

Fig. 5.19 Simulation setup of the system in study. The map of the frequency-dependent dielectric constant distribution for the inhomogeneous breast interior is shown at 1 GHz. The tumour is surrounded by fibroglandular tissue of low-adipose content. The system's seven transceivers are distributed along the shown ellipse, in [33]

method. A uniformly distributed elliptical array of seven point sources shown in Fig. 5.19 is considered for the construction of the MDM. Tumours are modelled as described in previous sections.

To examine the possibility of DORT-based tumour classification via contrast-enhanced differential backscatter, a gradual decrease in the tumour complex permittivity down to 70 % of its initial value is considered, in accordance with recent findings in [42]. It is also assumed that the decrease in the dielectric properties inside the tumour occurs homogeneously. As the tumour does not resemble a point-like target, it will generate multiple non-negligible singular values which depend on its shape and dielectric properties, as well as the polarisation of the electromagnetic wave and the properties of the background medium [6]. An example of the SVD distributions for the pre- and post-agents differential response for MISP and OMA targets of comparable size is shown in Fig. 5.20. Significant differences emerge in the SVD spectra of the MISP versus the OMA tumours, and these differences are consistent for several random realisations of the targets. In particular, the most significant difference is the proximity of the second (green) and third (red) higher singular values for MISP-type tumours relative to OMA tumours. This feature was observed for all random realisations of the two tumour categories, and is clearly visible with the dB scale shown in Fig. 5.20c, d, where the singular values have been normalised to their maximum value of the upper curve.

The SVD distribution is dependent on the size of the target size, as larger targets will generate more significant singular values. This can cause false positive results in the classification of tumours of unknown size. A possible solution to this problem is a two-stage classification, where the size of the tumour is estimated at the first stage, followed by a shape classification of tumours of comparable size [17]. Given that accurate differential data is obtainable by two successive scans of the breast before and after the infusion of the agents, the proposed approach may be a valuable tool for tumour classification. As with previous methods, measurement noise should not significantly affect these results, as long as the dynamic range of the system exceeds the noise level. The effect of pre-contrast and post-contrast measurement mismatch can be an important challenge for all methods based on differential imaging, which

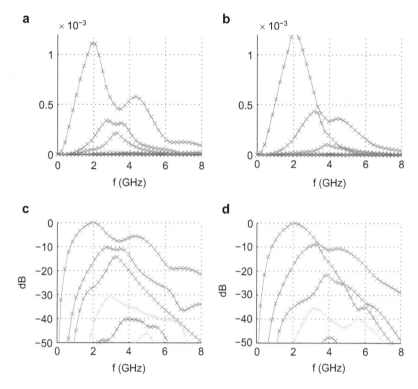

Fig. 5.20 Examples of SVD plots versus frequency for the pre- and post-agents differential target response in the case of (**a**) an MISP target, and (**b**) an OMA target. In (**c**), (**d**) the singular values are plotted in dB scale, in [33]

merits further investigation. In addition, measurement uncertainties such as small errors about the knowledge of the imaging apparatus setup or the position of the biological structure under monitoring may undermine the detection or classification of tumours via all the above methods. These and other considerations, such as possible alterations in the tumour morphology due to the accumulation of the contrast agents, must be studied further to fully assess the effectiveness of all these approaches for tumour classification.

5.3 Classification Based on Radar Target Signature Classification of Tumours

Studies by Davis et al. [17], Conceição et al. [12–14], McGinley et al. [44] and O'Halloran et al. [51] addressed the classification of breast tumour models based on their RTS obtained through ultra wideband microwave backscatter in 3-D simulations. Such studies were based on tumour models based on GRSs, as presented in Sect. 5.1.2.

The different GRS tumour models, as well as breast tissue, were modelled in a 3-D Total-Field/Scattered-Field (TF/SF) FDTD model with the dielectric Debye parameters for malignant tissue and for normal breast tissue [3, 35], respectively. Two main simulation scenarios are proposed: classification of tumours embedded in a homogeneous breast phantom [12, 13, 17, 44] and also in a breast phantom that accounts for some heterogeneity [12, 14, 51]. The RTS of tumours is then extracted from microwave backscatter and used to classify tumours, since the RTS contains valuable information about the shape and surface texture of tumours. The most important characteristics of any classification system are the Feature Extraction Method (FEM) used, along with the type of classifier itself. In this section, the performance of different FEMs in combination with different classifiers to classify breast tumours in terms of their size and shape is examined.

The tumour classification algorithm comprises two steps: the extraction of relevant features from each recorded backscattered signal and the classification method itself. The FEMs which have been considered in the literature for these studies include: Local Discriminant Basis (LDB), Principal Component Analysis (PCA), Independent Component Analysis (ICA) and Discrete Wavelet Transform (DWT). The classification methods which have been considered include LDA, QDA, SVM, SNNs and SOMs. Many of these classifiers have been used in several different multi-stage classification architectures. The following two sections examined the FEMs and classification algorithms.

5.3.1   Feature Extraction Methods

The backscattered signals are firstly processed through a feature extraction method so that the most relevant features from each signal can be highlighted and forwarded to the classification algorithms. The following FEMs have been considered in the literature: PCA, LDB, ICA and DWT.

5.3.1.1   Local Discriminant Basis

The LDB algorithm operates by selecting the 'best' basis from an overcomplete dictionary of time-frequency functions for classification [61]. In [17], a wavelet packet dictionary based on a third-order Daubechies wavelet was used to obtain coefficients for the training data. The atom in the wavelet packet, $\mathbf{h}_m$, in which the best basis vector is given by $\mathbf{m} = \{j, k, l\}$ and j is the jth scale, k is the kth frequency and l is the lth position [77]. The mth coefficient for the ith training vector from class c is given by $\theta_m^{(i,c)} = \mathbf{h}_m^T \mathbf{x}_i^{(c)}$. The best basis vectors $\mathbf{m}$ for classification is calculated through an additive discriminant measure D [61]. A time-frequency energy map for the training data is given by Saito and Coifman [61]

$$\Gamma_c(m) = \sum_i (\theta_m^{(i,c)})^2 / \sum_i \| \mathbf{x}_i^{(c)} \|^2 \qquad (5.12)$$

and the discriminant measure D—the symmetric Kullback–Leibler distance—which measures the divergence of energy distributions between two or more classes, is given by

$$D(\Gamma_{c_1}(m), \Gamma_{c_2}(m)) = \sum_l \left[\Gamma_{c_1}(m) \log \frac{\Gamma_{c_1}(m)}{\Gamma_{c_2}(m)} + \Gamma_{c_2}(m) \log \frac{\Gamma_{c_2}(m)}{\Gamma_{c_1}(m)} \right] \quad (5.13)$$

$$D\left(\{\Gamma_{c_n}(m)\}_{n=1}^C\right) = \sum_{r=1}^{C-1} \sum_{s=r+1}^{C} D(\Gamma_{c_r}(m), \Gamma_{c_s}(m)). \quad (5.14)$$

The summation in (5.14) is over the position index l implicit in the $\mathbf{m} = \{j, k, l\}$ wavelet packet index. The LDB is a fast algorithm which obtains the basis $\mathbf{H}$ with the best discriminant measure D within an overcomplete wavelet packet tree.

5.3.1.2 Principal Component Analysis

PCA allows for the reduction of the dimensionality of multivariate data, revealing simplified structures which are often hidden in the original data set while disregarding less relevant information, such as noise or collinearities in signals [70, 78]. This process is accomplished by applying a linear algebraic operation to the data set in such a way that the new orthonormal bases (i.e. the principal components) of the original data set present maximal variance, and so discrimination of different classes within the data set is easier to be completed by a classifier. The principal components are ordered by decreasing variance, and furthermore the variance along each principal component provides a measurement of the relative importance of each dimension [70]. In order to avoid computational complexity, the minimum number of principal components that is representative of the original data set should be assessed [1]. To obtain the principal components of a matrix X represented by $(m \times n)$, where m is the number of measurements and n is the number of samples, the mean of the sample for each ith measurement is subtracted and finally the basis vectors, $\mathbf{h}_m$, which are the eigenvectors of the covariance matrix $\mathbf{C} = E\left\{\widehat{\mathbf{X}}\widehat{\mathbf{X}}^T\right\}$, are calculated. The mean-corrected data are represented, for each ith measurement, by its Karhunen–Loéve expansion

$$\mathbf{X} - E\{\mathbf{X}\} = \sum_{m=1}^{N_m} \theta_m \mathbf{h}_m \quad (5.15)$$

in which θ_m represents each basis expansion coefficient and N_m represents the full dimensionality of the problem [17, 70].

5.3.1.3 Independent Component Analysis

Similarly to PCA, ICA is used in multivariate data sets in order to reduce the representation and capture the essential structure of the data. However, for ICA, statistically independent components are extracted through linear transformations from data with a non-Gaussian distribution, whereas for PCA it is assumed that the data have a Gaussian distribution [29]. To describe ICA, the following notation is defined: the observed data are given by x which stands for the random vector whose elements are the linear mixtures $x_1, \cdots, x_n$; the independent components are given by s which is the random vector with elements $s_1, \cdots, s_n$; and n stands for the number of independent components; A stands for the mixing matrix with elements a_{ij}, the parameters that relate x with s. The ICA model can then be defined as

$$x = As. \tag{5.16}$$

For the ICA algorithm, only the random vector x is directly observed, and so both A and s have to be estimated. ICA implies that the components s_i are statistically independent and also that they have non-Gaussian distributions. The calculation of s is given by

$$s = Wx \tag{5.17}$$

in which W is the inverse of A. Finally, W can be estimated by iteratively balancing a cost function which either maximises the non-Gaussianity of the calculated independent components or minimises the mutual information, in a process detailed in [29]. FastICA is a popular algorithm to calculate W, which is based on a fast fixed-point iteration scheme for maximising non-Gaussianity as a measure of statistical independence for ICA [22, 29]. Once W is estimated, the independent components can be obtained using (5.17).

Comparing ICA to PCA, it is worth noting that ICA has the following limitations: (1) the variance of the independent components cannot be determined, and (2) the order of the independent components cannot be determined [22, 29].

5.3.1.4 Discrete Wavelet Transform

The DWT produces wavelet coefficients which may be used as discriminant bases for classification methods [41, 61]. Wavelets are localised basis functions which are translated and dilated versions of a chosen fixed mother wavelet, the wavelet which will produce optimised classification for a certain data set. Wavelets allow the extraction of localised frequency information from a function or a signal. Wavelet families include: Daubechies, Coiflet, Symlet, Discrete Meyer, Biorthogonal and Reverse Biorthogonal [17, 40, 41, 61, 74].

When the DWT is applied to a set of RTSs, the wavelet coefficients are obtained by the decomposition low-pass filter and the decomposition high-pass filter. Subsequently, the low-pass band may be split again through a decomposition low-pass filter and a decomposition high-pass filter. It must be noted that for each iteration of the wavelet filters, the number of samples for the next stage is halved through signal decimation. This process continues to a desired number of levels. The final wavelet coefficients, given in a specific frequency sub-band, are supplied to the subsequent classification methods.

For the results presented in section, the chosen wavelet was Coiflet 5 [76], as this wavelet produced the best performance. Coiflet 5 was chosen from a comprehensive library of 126 wavelet functions [40, 76] after analysing preliminary classification results with LDA, which is described in detail in the following Sect. 5.3.2.1.

The difference between the DWT method used here, and the LDB method Sect. 5.3.1.1 is twofold: (1) a different wavelet was used, a fifth-order Coiflet wavelet for DWT, and (2) also up to four levels of decomposition were obtained to test the full library of wavelet coefficients.

5.3.2 Classification Algorithms

Five classification methods have been used in the literature to assess the size and shape of 3-D breast tumour models: LDA, QDA, SVM, SNN and SOMs and these are detailed in the following sections.

All these classification methods have been applied to the backscattered signals collected for each tumour model using the cross-validation method, so that the performance of each classifier is evaluated using a testing set independent from the training set, hence minimising the generalisation error, i.e. the ratio of misclassified samples [28]. The cross-validation method is used as follows. For A-fold cross-validation, the full set of backscattered signals is divided in A subsets, each of which contains one sample of each type and each size of tumours; each subset is then tested against the remaining $(A - 1)$ subsets, and, finally, all resulting A sub-classifications are averaged to obtain the overall accuracy classification performance.

For a subset of these classification algorithms, a set of up to eight multi-stage classification architectures, which categorise the data according to different levels of granularity in size or shape, have also been considered in this chapter; these are presented in Sect. 5.3.3.

5.3.2.1 Linear Discriminant Analysis

LDA is a technique that allows for discrimination of groups which have multivariate normal Gaussian distributions and have the same covariance matrix. For a discriminant analysis there are dependent variables which represent the classes of the objects and independent variables which are the object features that may describe each class. In case each object is defined by two features, the separator between two groups of objects is a line, otherwise if objects are defined by three or more features the separator is a hyperplane [20, 26, 34, 57, 67].

With the LDA method, a linear combination representing the weighted sum of two or more independent variables defines the discriminant function, which will allow classification [20, 26, 34, 57, 67]. The discriminant score for each object, k, in the analysis is a summation of the values obtained by multiplying each independent variable, X_{ik}, by its discriminant weight, W_i. The discriminant Z score for each object is given in the equation below

$$Z_j k = a + W_1 X_1 k + W_2 X_2 k + \ldots + W_n X_n k \tag{5.18}$$

in which a is the intercept and n is the total number of independent variables. In case there is more than two groups being discriminated at once, a series of classification functions is derived, which depends on the number of each pair of groups [26]. The mean, or centroid, for each group is obtained by averaging the discriminant Z scores for all objects within a group, and it represents the most typical location of an object from a particular group. There are as many centroids as the number of groups being classified at once. The distribution of the discriminant Z scores for each group influences the classification greatly [26]. For LDA, a linear separator is built based on the centroid and the distributions of the discriminant Z scores. Classifiers based on LDA can be applied to the data using the cross-validation method so that the classification performance is evaluated using a testing set, independent from the training set [28].

5.3.2.2 Quadratic Discriminant Analysis

QDA is a technique that allows for discrimination of classes with significantly different class-specific covariance matrices, while the class populations represent multivariate normal Gaussian distributions with the same mean [20, 34, 67]. For QDA there are dependent variables which represent the classes of the objects and independent variables which are the object features that may describe each class. As a result of having different covariance matrices for each class, the quadratic discriminant function involves both squared and cross-product terms. For a quadratic discriminant function, for a two-class scenario, an individual vector of scores x is classified as belonging to Group 1 if the following inequality holds:

$$\mu_1'(\Sigma_2^{-1} - \Sigma_1^{-1})x - 2x'(\Sigma_2^{-1}\mu_2 - \Sigma_1^{-1}\mu_1) + (\mu_2'\Sigma_2^{-1}\mu_2 - \mu_1'\Sigma_1^{-1}\mu_1) \geq$$

$$ln\frac{|\Sigma_1|}{|\Sigma_2|} + 2ln\frac{|\pi_1|}{|\pi_2|}$$

$$\tag{5.19}$$

in which the class means are given by μ_1 and μ_2, and the covariance matrices are given by Σ_1 and Σ_2, for Groups 1 and 2, respectively, π_1 and π_2 are the prior probabilities of observing a member of Groups 1 and 2, respectively [26, 57]. Generally, QDA offers increased flexibility over LDA at the cost of possibly overfitting the training data [20]. Similarly to LDA, classifiers based on QDA can be applied to the data using the cross-validation method so that the performance of each classifier is evaluated using a testing set, independent from the training set [28].

5.3.2.3 Support Vector Machines

The SVM machine learning algorithm is typically used as a method to handle nonlinear relations between the input vectors and their corresponding labels by transforming linearly inseparable data to a higher-dimensional space in which they can be more readily separated, usually into two groups [2, 4, 5, 16, 28, 49]. For the particular case of the SVM, the input vectors are mapped to a higher-dimension feature space by means of a Kernel (K) [2, 5].

The Kernel used for this study is the radial basis function (RBF), which allows for all input vectors to be nonlinearly mapped in an infinite-dimension feature space, typically a Hilbert space. The decision hyperplane can then be obtained in the feature space and is generically given in the following format:

$$\mathbf{w}x + b = 0 \tag{5.20}$$

in which $\mathbf{w}$ is the normal to the hyperplane, x represents the data and b is the bias.

Knowing that the data can be represented by the inner product $x_i \cdot x_j$ (this is an implication of using an infinite feature space), the equation for the RBF is defined as follows:

$$K(x_i, x_j) = \exp(-\gamma \left\| x_i - x_j \right\|^2), \ \gamma > 0 \tag{5.21}$$

in which γ is the scaling factor of the RBF Kernel [4, 28].

The decision hyperplane is supported by two parallel vectors, one on each side of the hyperplane. Each of these support vectors is at the same distance from the hyperplane (the 'margin') and each of them delimits either the first or the second labelled class. A classifier will work better when the value for the margin is maximised, so the concept of a soft margin is introduced, as opposed to hard margins, as described in [2, 4, 5]. When soft margins are used, it implies that the support vectors are most likely built with supporting samples that represent meaningful samples of the training group, while outlier samples, such as noisy data or unusual data, are ignored for the calculation of the support vectors. If such conditions are met, the learning machine ensures high generalisation [16, 28, 49] and therefore will be able to successfully classify an independent testing group. Knowing that the training set is composed of sample-label pairs (x_i, y_i), in which $i = 1, \ldots, l$ represents each sample, x_i represents the input vectors of each sample and y_i represents the respective label, soft margins can be calculated by following the mathematical optimisation

$$\min_{w,b,\xi} \left[\frac{1}{2} (\mathbf{w} \cdot \mathbf{w}) + C \sum_{i=1}^{l} \xi_i \right] \tag{5.22}$$

with the following conditions: $y_i(\mathbf{w} \cdot x_i + b) \geq 1 - \xi_i$, in which the slack variable $\xi_i \geq 1$. For a hard margin the data are scaled so that the margin equals 1, while for a soft margin the margin can be below one as it is given by $1 - \xi_i$. However, this results in the increase of the objective function since the sum of errors, given by $\sum_{i=1}^{l} \xi_i$, is multiplied by C [5, 28]. The function of C is two-fold: it controls the relative weighting to keep $\mathbf{w} \cdot \mathbf{w}$ small (as the size of the margin is maximised), and it ensures that most samples have a functional margin of at least one [49].

For application of SVM, the data need to be pre-processed so that the SVM classification algorithm can be optimised adequately for the samples, as efficiently as possible. These pre-processing steps are as follows:

1. Scaling of the training and data set;
2. Application of the Kernel function, RBF;
3. Application of the Cross-Validation method;
4. Optimisation of the RBF parameters.

In the first step, the input vectors (represented by the signals pre-processed with a FEM) for each sample in both training and testing groups are scaled to the range $[-1, +1]$. This step is important so that attributes in greater numeric ranges do not dominate those in smaller numeric ranges, and also that the computational load of the whole algorithm is restricted [28].

For the second step of the SVM algorithm, the RBF is applied.

Another issue that has to be considered is the fact that the combinations of (C, γ) for the RBF Kernel have to be tested on the training data through a cross-validation method. This method allows for outlier samples that represent noise or unusual data to be removed, and as a result, some of the outlier supporting samples may be omitted from the final solution.

Finally, the parameters of the chosen Kernel function are adjusted so that the classifier is successful in classifying independent testing groups. For the RBF, the combination of (C, γ) is optimised, C is the penalty parameter of the error term, and as earlier mentioned, γ is the scaling factor of the RBF Kernel. A parameter search, such as the grid-search described in [28], is applied to the data set.

5.3.2.4   Spiking Neural Networks

SNNs are more closely related to their biological counterparts than previous artificial neural networks generations, such as multi-layer perceptrons. SNNs, in contrast to previous models, employ transient pulses for communication and computation. Maass has demonstrated that spiking neurons are more computationally powerful than threshold-based neuron models [38] and that SNNs possess similar and often more computation ability compared to multi-layer perceptrons [39].

Inspired by nature, a genetic algorithm (GA) [27] models natural evolution through a set of computational operators. A GA is a parallel, population-based search strategy that encodes individual solutions into a data-structure known as a genome. A population of such genomes is maintained by the GA and mechanisms analogous to evolution are employed to evolve high-fitness solutions. Exploration of the search space is performed using a diversity introducing mutation operator while

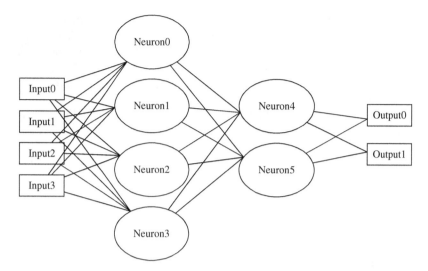

Fig. 5.21 Example fixed-topology spiking neural network. Images reproduced courtesy of The Electromagnetics Academy [50]

crossover (mating of two parent solutions) is employed to exploit good solution building blocks (known as genes) already in the population. Selection pressure is added through a tournament selection operator to incorporate 'survival of the fittest'. Traditionally, SNN simulations involve constructing a fixed, regular structure of neurons arranged in layers where all neurons are fully forward connected, as shown in Fig. 5.21. This approach simplifies the design of the network and provides a structure whose neuron firing thresholds and synaptic interconnect weights may be evolved, with a fixed-length genome, to form a solution. Recently, there has been growing interest in exploiting the adaptability provided by GAs to modify the interconnect structure of an SNN in order to create a topology which is both simpler and more suitable for the task at hand. A common issue with this concept is designing an appropriate encoding mechanism for the structure of the network such that it may be mutated and combined with other networks in a feasible manner. Additionally, network structures evolved by a GA have a tendency to grow as the GA progresses. This topology growth results in a more complex search space partly negating the advantage of evolving a task specific network.

The NeuroEvolution through Augmenting Topologies (NEAT) [72] algorithm which is tailored to address these particular concerns was incorporated. The NEAT algorithm incorporates historical markers in the SNN gene which allows genes with common ancestors to be combined as part of the crossover mechanism of the GA. NEAT also uses this historical information to group individuals into species based on common ancestors [72]. When the GA creates a new generation, the selection of individuals (i.e. choosing which individuals will be combined together to form an individual for the new population) is traditionally based on each individual's fitness. NEAT implements explicit fitness sharing [24] within species, where individuals in a species must share their combined fitness (i.e. the fitness of an individual is modified to be the average fitness of all individuals in that species). This deters species from growing too large, as each individual must contribute to the species fitness, hence allowing many diverse species (i.e. many unique approaches to solving the problem) to co-exist. Species whose fitness does not increase over a number of generations become extinct (i.e. the individuals are deleted from the population and replaced with new, randomly initialised, individuals) ensuring individuals continue to improve as the network complexity grows. Traditional fully connected SNNs will contain many neurons and connections which do not contribute to classifier accuracy. The NEAT operators create networks which are of optimal size and only contain neurons and connections which aid the function of the classifier.

Pre-Processing and Fitness Function

The classifier considered here is a two-class problem (i.e. malignant versus benign). The DWT is applied to extract the most significant classification features of the RTS, in a process previously described more comprehensively in [51]. In this study, large DWT values are mapped to high spike frequencies while small DWT values are mapped to low frequencies. Since values are scaled between $[-1; +1]$, it is necessary to decouple the positive and negative ranges of each DWT component $(D(n))$ into two spike generating inputs $(D(n)+$ and $D(n)-)$. This decoupling ensures that a $+1$ DWT input generates the same number of spikes (and influence) on the SNN as a -1 DWT input, thus removing any bias from the encoding process. The SNN processes the 15 most significant DWT components. Thirty spike generators are used to map real-valued DWT data into spike trains using a linear magnitude to (spike train) frequency conversion [54].

Two output layer spiking neurons generate two spike trains, which are processed by two spike counters to count the number of output spikes within a given update interval [54]. These counter values are used to determine classifier behaviour. The counter with the largest spike count value designates the selected class. The neuron model chosen for these experiments is based on the leaky integrate and fire model [38]. Each SNN individual is initially composed of thirty input neurons and two output neurons. The NEAT GA progressively adds neurons and connections and hence each individual has a variable number of genes. The GA also modifies the weights on the synaptic connections and neuron firing threshold.

Synaptic weights range from $[-1; 1]$ while thresholds vary between $[0; 5.0]$ [59]. Fitness assessment of the SNN-based breast cancer classifier is achieved using a fitness function, which rewards individuals based on the number of correct classifications made. Cm refers to the number of correct malignant classifications made by the SNN. Cb refers to the number of correct benign classifications. $Cmax$ and $Cmin$ are defined in Eqs. 5.24 and 5.25. The fitness function, f, of the SNN is defined as follows:

$$f = C_{\min}\beta + C_{\max} \tag{5.23}$$

where:

$$C_{\max} = \max(C_m, C_b) \tag{5.24}$$

$$C_{\min} = \min(C_m, C_b) \tag{5.25}$$

A β value of 1.6, chosen through empirical analysis, is employed in this research to reward the correct classification of both tumour classes. Without this fitness bias, fitness can be accumulated by classifying a single tumour class repeatedly. By including a β value greater than one, networks that select correctly from both classes are rewarded above networks that correctly select from just one class.

5.3.2.5 Self-Organising Maps

SOMs are a type of neural network that are trained using unsupervised learning, where the input pattern is applied and the network produces the output without being told what output should be produced [31]. SOMs consist of an input and an output layer. The topology of a SOM network is shown in Fig. 5.22. The dimension of the input layer is defined as being equal to the number of features or attributes, while the output layer is typically a two-dimensional grid (shown as red, white and yellow regions in Fig. 5.22). In SOMs, the two layers are fully interconnected, i.e. each input (ip_i) is connected to every unit or node in the output layer.

To illustrate the operation of the SOM, two-dimensional data is employed here, showing the topological mapping of the data. Although, 2-D data is used here for illustration purposes, the SOM performs very well in organising much higher-dimensional data.

Fig. 5.22 Kohonen SOM topology, adapted from [60]. Each output layer node is represented by an N-dimensional weights vector, in [30]

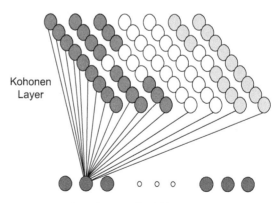

Each Output Node is a vector of N weights

Kohonen Layer

Input Layer – Each Node a vector representing N terms.

Fig. 5.23 Plot of network
weights directly after random
initialisation, in [30]

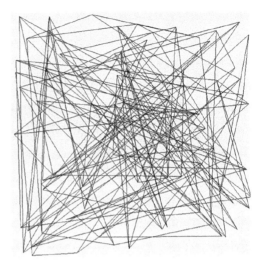

The 2-D input data is randomly initialised and evenly distributed over the range
from zero to one. Weightings, wt_{j1} and wt_{j2}, initially also randomly selected from
the same range, are associated with the inputs to each node j. These weights are
adapted so that the network of weights, as an entirety, organises to form topological
mappings of the input space. This means that the distribution of weight values in
the network will reflect the distribution of input data. To more easily visualise the
topological distribution of network weights, a graph is plotted with a point for each
node in the output layer, the coordinates of each point being given by the weight
values of the node (e.g. ordinate value wt_{j1} and abscissa value wt_{j2}). If nodes in the
output layer are assigned indices (i, j) denoting their row and column, then joining
the point for node (i, j) to the points for nodes $(i + 1, j)$ and $(i, j + 1)$ for every
node in the output layer yields a plot similar to Fig. 5.23. Figure 5.23 illustrates
randomly selected weights chosen from the range zero to one before the Kohonen
training process was applied (i.e. the plot was made directly after initialising the
weight values).

Network Training: Having randomly initialised the weight values, the training
process described by Kohonen [31] now begins. For training, the following steps
are repeated for T iterations, where T is the number of training steps.

1. Randomly choose inputs to present to the SOM;
2. On the basis of a Euclidean distance metric, find the output layer node whose
   weights are most similar to the input;
3. Update the weight of that node and those of its neighbours according to the
   following equations:

$$\delta wt_{ji} = \alpha(ip_i - wt_{ji}) \tag{5.26}$$

$$wt_{ji\text{new}} = wt_{ji} + \delta wt_{ji} \tag{5.27}$$

Fig. 5.24 Network weights
after training has taken place,
in [30]

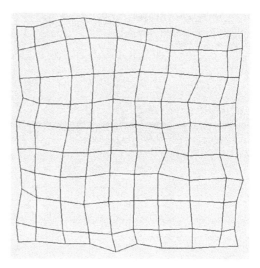

where wt_{ji} is the weighting between node j and input i, ip_i is the ith input, and α is the gain or learning rate (an empirically chosen adjustable parameter that can be adjusted to regulate the training speed);

4. Reduce neighbourhood size and learning rate as per the following two equations:

$$D = \lceil d_0(1 - t/T) \rceil \qquad (5.28)$$

where d_0 is the initial neighbourhood size, t is the current updating/training iteration and T is the total number of iterations.

$$\alpha = \alpha_0(1 - t/T). \qquad (5.29)$$

Typically the neighbourhood size begins large (e.g. one-half to one-third of the grid size). Several different forms of neighbourhood type can be used. In this section, a simple square neighbourhood is used and neighbourhood size is restricted to integer values. After training has been completed, the weight values are once again plotted and shown in Fig. 5.24.

Network Testing Randomly chosen input patterns are applied and Euclidean distance competitions held to see which set of weights are most similar to the input patterns. It is found that similar inputs pattern cause nodes that are adjacent in the output layer to win. This being the case, from Kohonen's definition [18], the neural network can be said to be organised: 'The mapping is said to be ordered if the topological relations of the images and the patterns are similar'. Using a skewed input distribution, where the second training input is chosen to be in the range [0;0.2] when the first training input is greater than 0.5, leads to the map shown in Fig. 5.25.

This approach to testing and training, applied to the breast tumour RTS data set, is discussed in Sect. 5.3.4.

Fig. 5.25 Plot of network
weights for a skewed input
distribution, in [30]

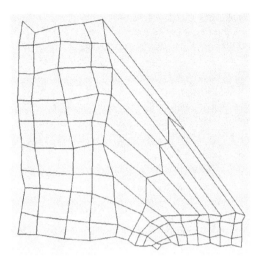

5.3.3   Classification Architectures

Eight different classification architectures are considered, six of which are composed of a number of binary sub-classifiers (and are studied with LDA, QDA and SVM), and the remaining two are composed of direct sub-classifiers that classify four classes at once (and are used with LDA and QDA). The different architectures are defined by the size and shape granularity, i.e. by the number of categories classified in each stage (two or four categories), and by the number of stages each size and/or shape classifier is composed of (one or two stages). A coarse-shape classifier is used to classify tumours into either malignant or benign tumours. However, extra granularity in the shape classifier allows further classification of tumours into spiculated, microlobulated (both malignant tumours), and in macrolobulated and smooth (both benign tumours), giving additional clinical information on the development stage of a breast tumour. For instance, a macrolobulated shape could potentially be an indicator of pre-malignancy.

Each of the classifier architectures will now be described. The first classifier architecture, Coarse-Shape (CS), splits the RTS of the tumours in one stage into two shape names: malignant or benign. Similarly, the Fine-Shape (FS) initially classifies the RTS of the tumours into the same shape categories as the CS, but then adds another level of shape granularity by dividing malignant tumours into spiculated and microlobulated tumours, and benign tumours into macrolobulated and smooth tumours. The CS and FS architectures can be used with LDA, QDA and SVM classifiers and are both shown in Fig. 5.26.

The Coarse-Size-Coarse-Shape (CSCS) splits the RTS of the tumours in one stage into two size groups (the first group contains smaller 2.5 and 5 mm tumours and the second group contains bigger 7.5 and 10 mm tumours), before further classifying the tumours into either benign or malignant. Similarly, the Coarse-Size-Fine-Shape (CSFS) initially classifies the RTS of the tumours into the same size

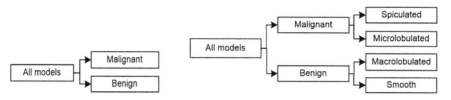

Fig. 5.26 Classification architecture in which only shape classifications are applied: CS (*left*) and FS (*right*), in [11]

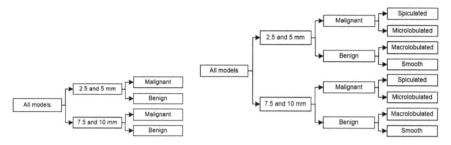

Fig. 5.27 Classification architectures in which a 1-stage coarse size classification is applied before shape classification: CSCS (*left*) and CSFS (*right*), in [11]

and shape categories as the CSCS, but then adds another level of shape granularity by dividing malignant tumours into spiculated and microlobulated tumours, and benign tumours into macrolobulated and smooth tumours in a second stage of shape classification. The CSCS and CSFS architectures can be used with LDA, QDA and SVM classifiers and are both shown in Fig. 5.27.

The Fine-Size-Coarse-Shape (FSCS) and Fine-Size-Fine-Shape (FSFS) classification architectures further classify the RTS of the tumours into four subcategories of size (2.5, 5, 7.5 and 10 mm) in two stages. The FSCS then divides them into two categories of shape, benign and malignant, while the FSFS classifies them into four shape categories: spiculated, microlobulated, macrolobulated and smooth in two stages. The FSCS and FSFS classifiers can be used with LDA, QDA and SVM classifiers and are shown in Fig. 5.28.

5.3.4 Classification Results

Many reported studies in the literature have used different combinations of FEMs (such as those in Sects. 5.3.1.1, 5.3.1.2, 5.3.1.3 and 5.3.1.4) and classification algorithms (such as those in Sects. 5.3.2.1, 5.3.2.2, 5.3.2.3, 5.3.2.4 and 5.3.2.5), as well as different classification architectures Sect. 5.3.3. Results from different combinations of FEMs and classification algorithms are presented in the following sections.

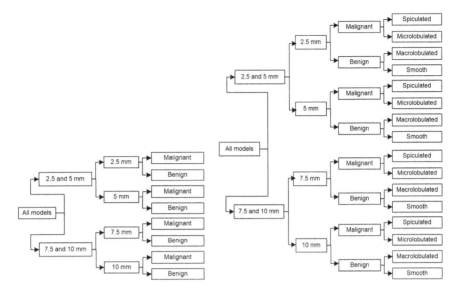

Fig. 5.28 Classification architectures in which a 2-stage fine size classification is applied before shape classification: FSCS (*left*) and FSFS (*right*). Images in [11] and reproduced courtesy of The Electromagnetics Academy [14]

5.3.4.1 Classification Results using PCA, ICA and DWT with LDA, QDA and SVM

Results in Homogeneous Breast Model

In this section, a database of 480 GRSs, such as those presented in Sect. 5.1.2, is used with GRSs of 4 different sizes and 4 different shapes. Firstly, different FEMs are used to extract the most significant bases from the RTS of the tumours: PCA, ICA and DWT (with Coiflet 5 wavelet). In particular, a dimensionality reduction of PCA is accomplished as a pre-processing step so that the more representative bases of the data are selected—the selection of 30 principal components offers a good compromise between classification accuracy and computational time. Regarding ICA, the FastICA algorithm outputs 16 independent components to represent each signal. For the DWT method, the wavelet function in use is Coiflet 5, and the wavelet coefficients from the low-pass sub-band after the decomposition of the low-pass filter with two iterations represent the best features for classification. Further detail is given in [11].

SVM is applied to the six binary-based classification architectures described in Sect. 5.3.3 (CS, FS, CSCS, CSFS, FSCS and FSFS). The performance for the SVM-based classification architectures depends on optimal values for parameters γ and C, as described in Sect. 5.3.2.3. The downsampled UWB backscattered signals are processed through the same three FEMs: PCA, ICA and DWT, and the SVM parameters are optimised for best classification performance with each feature,

Table 5.1 Classification accuracies (%) for size and subsequent shape classifiers and overall size-then-shape classifier using support vector machine (SVM) binary classifier for six different classification architectures (Coarse-Shape, CS; Fine-Shape, FS; Coarse-Size-Coarse-Shape, CSCS; Coarse-Size-Fine-Shape, CSFS; Fine-Size-Coarse-Shape, FSCS; Fine-Size-Fine-Shape, FSFS)

Architecture	Method	Partial size	Partial shape	Size-then-shape
CS	PCA	N/A	*89.20*	N/A
	ICA	N/A	84.66	N/A
	DWT	N/A	86.08	N/A
FS	PCA	N/A	*72.73*	N/A
	ICA	N/A	66.19	N/A
	DWT	N/A	67.33	N/A
CSCS	PCA	*94.89*	90.62	85.99
	ICA	94.32	87.21	82.26
	DWT	94.60	*91.19*	*86.27*
CSFS	PCA	*94.89*	*75.00*	*71.16*
	ICA	94.32	70.17	66.18
	DWT	94.60	74.43	70.41
FSCS	PCA	*86.93*	90.34	*78.53*
	ICA	79.83	86.93	69.40
	DWT	84.38	*90.91*	76.70
FSFS	PCA	*86.93*	*75.28*	*65.44*
	ICA	79.83	69.60	55.56
	DWT	84.38	75.00	63.28

The Feature Extraction Methods include: Principal Component Analysis (PCA), Independent Component Analysis (ICA), and Discrete Wavelet Transform (DWT). The best accuracy for each stage of the classification architecture is highlighted in *italic*. Table in [11] and reproduced courtesy of The Electromagnetics Academy [13]

through a grid-search procedure. For the results presented here, the optimal SVM parameters are as follows: for PCA: $\gamma = 23$ and $C = 28$, for ICA: $\gamma = 25$ and $C = 225$ and for DWT: $\gamma = 27$ and $C = 227$. Results for the different binary-based classification architectures are shown in Table 5.1, in which the best performance for each stage of the classification architectures is highlighted.

After a first analysis of Table 5.1 it is reasonable to assume that the FEM PCA allows for better classification using SVM in most cases when compared to the alternative features. By examining the six architectures in Table 5.1, it can be observed that applying a size classifier, both coarse and fine results in better performance than applying a shape classifier in isolation, also for both coarse and fine classifiers. Given these results, it is sensible to investigate the performance of a shape classifier when a size classifier is previously applied, like in architectures CSCS, CSFS, FSCS and FSFS.

Table 5.2 Classification accuracies (%) for size and subsequent shape classifiers and overall size-then-shape classifier using linear discriminant analysis (LDA), quadratic discriminant analysis (QDA) and support vector machines (SVM) binary classification algorithms for six different architectures of classifiers (Coarse-Shape, CS; Fine-Shape, FS; Coarse-Size-Coarse-Shape, CSCS; Coarse-Size-Fine-Shape, CSFS; Fine-Size-Coarse-Shape, FSCS; Fine-Size-Fine-Shape, FSFS), with γ set to 16 and C set to 256

Architecture	Partial size			Partial shape			Size-then-shape		
	LDA	QDA	SVM	LDA	QDA	SVM	LDA	QDA	SVM
CS	N/A	N/A	N/A	80.90	84.03	*89.20*	N/A	N/A	N/A
FS	N/A	N/A	N/A	58.33	55.90	*72.73*	N/A	N/A	N/A
CSCS	93.05	91.32	*94.89*	87.15	82.29	90.62	81.10	75.13	*85.99*
CSFS	93.05	91.32	*94.89*	67.36	62.15	*75.00*	62.68	56.76	*71.16*
FSCS	79.86	72.22	*86.93*	86.80	84.72	90.34	69.32	61.19	*78.53*
FSFS	79.86	72.22	*86.93*	69.44	64.58	75.28	55.46	46.64	*65.44*

Tables reproduced courtesy of The Electromagnetics Academy [13]
All results have 4 significant digits

In Table 5.2, the same database of 480 GRSs is tested with PCA as the FEM, and three classification algorithms: LDA, QDA and SVM.

For both Tables 5.1 and 5.2, in terms of the overall performance of the classifiers, it is observed that the more stages of partial size and/or shape classifiers the lower the overall performance. Conversely, the fewer stages of partial size and/or shape classifiers, the higher the overall performance. The performance of the classifiers decreases with the increase of granularity for two specific reasons: Firstly, for fine classification (e.g. differentiating between smooth and macrolobulated), the RTS of the tumours is quite similar, so classification is much more difficult and misclassification is more likely to occur. Secondly, when classifiers are grouped in architectures such as the ones used in this study, errors can propagate through the multi-stage classifier. For instance, a microlobulated tumour which is first classified as benign will never be classified correctly in a fine shape classifier (as it will automatically be misclassified as a smooth or a macrolobulated tumour).

For analysis purposes, errors in classification in terms of both size and shape were recorded at each stage of the classification architectures. In general it was observed that the number of tumours misclassified for one class was very similar to the number of tumours misclassified for the second class. However, there were two exceptions:

- The fine shape classifier misclassified several smooth tumours as macrolobulated and vice-versa the size pre-classification. This is due to the similarity between these two types of benign tumours.
- There was a significant number of spiculated tumours misclassified as microlobulated and vice-versa for larger tumours (with 7.5 and 10 mm radius). In smaller tumours models, the spicules extend further beyond the surface of the tumour

compared to larger tumours. In larger malignant tumours, the spicules may not influence the RTS of the tumours as much as for smaller spiculated tumours, and therefore misclassifications between larger spiculated and microlobulated tumours are more likely to occur.

Results in Heterogeneous Breast Model

A subset of 160 GRSs were tested in the following four different dielectrically heterogeneous breast scenarios: (I) breast model with a cluster of fibroglandular breast tissue in a fixed location independent of the tumour location; (II) breast model with a cluster of fibroglandular breast tissue, in a fixed location, possibly overlapping with tumour; (III) breast model in which one cluster of fibroglandular breast tissue is modelled across a range of random different locations within the breast; and finally, (IV) breast model in which two clusters of fibroglandular tissue are modelled across a range of random different locations within the breast.

By observing Table 5.3, it is observed that the introduction of a portion of fibroglandular breast tissue at a fixed location within the breast model (models I and II) does not significantly degrade the classification performance, suggesting that the algorithm is efficient under these specific conditions. Table 5.3 shows that a performance decrease is observed when classifying the shape of tumours embedded in breast models with varying locations of fibroglandular tissue, in Model III. It is also observed that there is further shape classification performance decrease when the number of fibroglandular clusters increases to two (Model IV).

Table 5.3 Classification accuracies (%) for size and subsequent shape classifiers and overall size-then-shape classifier using principal component analysis (PCA) and support vector machine (SVM) binary classifier for two different classification architectures

Architecture	Model	Partial size	Partial shape	Size-then-shape
FSCS	I	85.83	92.71	79.57
	II	85.62	91.25	78.13
	III	83.12	90.62	75.33
	IV	80.00	85.00	68.00
FSFS	I	85.83	73.96	63.48
	II	85.62	75.42	64.57
	III	83.12	68.12	56.63
	IV	80.00	61.25	49.00

Model I refers to breast models containing one piece of fibroglandular tissue at a fixed location, Model II refers to breast models with two pieces of fibroglandular tissue at fixed locations, Model III refers to breast models containing one piece of fibroglandular tissue at varying locations, Model IV refers to breast models with two pieces of fibroglandular tissue at varying locations. Tables reproduced courtesy of The Electromagnetics Academy [14]

5.3.4.2 Classification Results Using DWT with LDA and SNN

Results in Homogeneous Breast Model

A total of 160 tumour models were considered (80 tumours of size 2.5 mm and 80 tumours of size 7.5 mm). Within that group, there were 80 type 1 tumours (malignant), 40 type 2 tumours (macrolobulated benign) and 40 type 3 tumours (smooth benign). Two different classifier architectures are considered:

- A direct 'type' classifier that simply classifies each tumour as either benign or malignant;
- A two-stage classifier that classifies each tumour as either small or large, before classifying the tumour as either benign or malignant.

The tumour backscatter is classified using the SNN, but also using LDA, providing a useful baseline when examining the performance and robustness of the SNN classifier. In order to evaluate both classification methods, the entire data set is randomly shuffled and divided into 75 % (120 Tumours) and 25 % (40 Tumours) training and test groups, respectively. The classification process is repeated 10 times and the average performance of each classifier is calculated. The results are presented in Table 5.4. Across all tests (dielectrically homogeneous and heterogeneous), the SNN was shown to equal or significantly outperform the LDA classifier. The SNN was shown to provide 100 % classification accuracy using the two-stage classifier when the tumours were first pre-classified by size.

Results in Heterogeneous Breast Model

In order to examine the effect of increasing dielectric heterogeneity on the performance of the SNN classifier, two specific scenarios are considered (Models I and II defined above). In the first instance, a single piece of fibroglandular tissue surrounds the tumour, while in the second more difficult scenario two separate pieces of fibroglandular tissue are located around the tumour. The performance of the classifier in an increasingly heterogeneous environment is shown in Table 5.5.

Table 5.4 Classification accuracies (%) of linear discriminant analysis (LDA) and spiking neural networks (SNNs) classifiers in homogeneous breast model

Classifier	CS	CSCS	FSCS
LDA	73	74.34	93.48
SNN	73	100	100

The following architectures were considered: Coarse-Shape (CS), Coarse-Size-Coarse-Shape (CSCS), Fine-Size Coarse-Shape (FSCS)
Tables reproduced courtesy of The Electromagnetics Academy [51]

Table 5.5 Effects of dielectric heterogeneity on performance of SNN classifier

Model	CS	CSCS	FSCS
I	78	100	91.0
II	68	100	86.75

Model I refers to breast models containing one piece of fibroglandular tissue, while Model II refers to models with two pieces of fibroglandular tissue. The following architectures were considered: Coarse-Shape (CS), Coarse-Size-Coarse-Shape (CSCS), Fine-Size-Coarse-Shape (FSCS). Tables reproduced courtesy of The Electromagnetics Academy [51]

The performance of the SNN classifier drops by 10 and 4.25 % for one-stage type and pre-size classified large tumours, respectively, as heterogeneity increases. Overall, the SNN classifier is shown to be relatively robust to significant increases in dielectric heterogeneity. In fact, in the most dielectrically heterogeneous models (Model II), the average performance (across large and small tumours) of the two-stage SNN classifier was over 86 %.

The performance of an SNN classifier in a dielectrically heterogeneous breast was examined. The SNN was shown to significantly outperform the LDA classifier in the dielectrically heterogeneous Models I and II (data not shown). The SNN classifier was also shown to be relatively robust to increasing levels of heterogeneity within the breast. It must be noted that the performance and robustness of the classifier may partly be attributed to the fact that the tumour is positioned at a fixed location at the centre of the breast. Because the tumour position is fixed across all models, the classifier can more easily isolate the portion of the RTS influenced by the shape and size of the tumour and effectively disregard any surrounding heterogeneity.

5.3.4.3 Classification Results Using DWT with SOM

Two distinct data sets are considered: data from simulations where the tumour is located in homogeneous breast tissue and a second set where fibroglandular tissue is present. Each of these data sets consists of 360 tumour models of average size 2.5 mm (120 macrolobulated benign tumours and 240 spiculated malignant tumours, of which 120 had 3 spicules and 120 had 10 spicules) each comprising 15 normalised and scaled DWT coefficients. In order to evaluate the classifier, each data set is randomly shuffled and divided into ten combinations of 276 training and 84 testing tumours. The classification process is repeated 10 times, for each of the ten shuffled combinations of data (or shuffled files), and the average performance of the classifier is calculated.

For this particular study, weights are randomly initialised to be in the range [−1, 1] and training consists of repeatedly applying scaled patterns of the sets of

15 DWT coefficients randomly chosen from the 276 tumour models in the training sample, until the network organises itself. The output layer consists of a square 10×10 grid. The gain/learning rate, neighbourhood size and number of training steps are empirically chosen. During training, the neural network has no input indicating its class (independent variable), i.e. indicating whether the input pattern being presented to it belongs to a malignant or to a benign tumour model.

At the end of training, the network weights are frozen. At this point, a Euclidean distance competition is held for each node in the output layer, for each of the 276 training tumour models in the training set. The tumour model that has an input pattern most similar to the weights of a specific node will be assigned to that node. Only at this point is the training data set examined to discover which inputs correspond to malignant/benign tumour models.

Two-Way Tumour Classification

Two-dimensional 'Tumour Tracking Map' can be produced, as shown in Fig. 5.29. In these examples, the tumour models are classified into two discrete groups: benign and malignant (shown as the green and red regions, respectively). The shape of the various regions (e.g. benign or malignant) is largely irrelevant. The goal is to simply create a clear division between the groups within the data.

A random number generator is used to produce this 'Tumour Tracking Map', as a result running the program repeatedly means that different parts of the random number sequence are utilised. Consequently, different 'Tumour Tracking Maps' will be created, even though the same data set is being used for training. The 'Tumour Tracking Maps' shown in Fig. 5.29 are only two of many produced by the SOM method. Note that individual maps are reproducible when the same part of the random number sequence is used.

Three-Way Tumour Classification

Three-way classification is shown in Fig. 5.30. In here, benign, 3-spiculated malignant and 10-spiculated malignant tumours are represented as green, orange and red regions, respectively.

Fig. 5.29 Simple two-way benign-malignant classification using a SOM. The benign region is shown in *green*, while the malignant region is shown in *red*

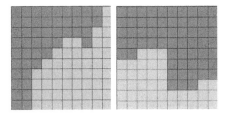

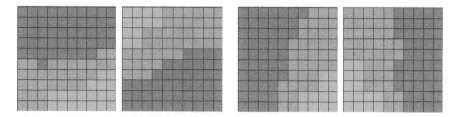

Fig. 5.30 Four examples of three-way classification between benign macrolobulated (*green*), 3 spicules malignant (*orange*) and 10 spicules malignant (*red*) tumours using a SOM. In [30]

In order to evaluate the performance of the three-way SOM classifier, the SOM is tested using a test sample of 84 tumours. Importantly, the DWT coefficients corresponding to these tumours have not previously been presented to the network. For testing, the 15 DWT coefficients for each of the tested tumour models are input to the trained network (whose weights are now frozen), a distance competition is held for each tumour model and a winning node found. The tumour is then classified based on which region it falls into on the SOM.

An average classification accuracy across 10 maps for each of the ten shuffled files was as follows:

- 99.5 % for macrolobulated tumours
- 90.54 % for 3 spiculated malignant tumours
- 88.28 % for 10 spiculated malignant tumours

yielding an overall average accuracy of 92.77 %.

The tumours fall into clear and distinct regions within the resultant SOMs. Significantly, the region of benign tumours (shown in green) is completely separated in all cases from the highly malignant 10-spicule tumour region (shown in red) by the intermediate 3-spicule tumour region (shown in orange). This suggests that the SOMs could potentially be used to monitor tumour development (similarly to when they were used to track bankruptcy in Spanish companies in [68]). This will be discussed in more detail in the following section.

Tumour Development Tracking using Self Organising Maps

Figure 5.31 illustrates the potential of a SOM to track tumour development. Here, the 15 DWT coefficients for one sample test tumour model (shown in blue), at three different stages of development, are applied to the trained SOM. When the tumour is macrolobulated benign, the winning node is found to be in the green macrolobulated region; when the tumour becomes more malignant, developing 3 spicules, it falls into the orange region; finally when the tumour becomes highly malignant (10 spicules), it is found to be in the red part of the map, as expected.

Fig. 5.31 Tumour
development tracking with a
SOM. As the tumour
becomes increasingly
malignant, it moves across
the map from the *green*
region to the *red* region.
Adapted from [30]

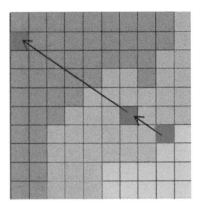

Three-Way Tumour Classification in Heterogeneous Breast Tissue

The performance of the classification algorithm was also examined in a more representative dielectrically heterogeneous breast tissue. A single piece of glandular tissue is placed close to the tumour at an average distance of 20 mm. Therefore, the resultant RTSs are composed of reflections from both the tumour and the surrounding glandular tissue, making the classification problem more challenging and realistic. Two classification approaches are considered: firstly the tumours are classified by using the procedure described when the RTS recorded at a single antenna at a time is used to classify the tumour ('one-view classification' detailed previously in [30]); in the second approach, the RTS from all four antennas surrounding the tumour are used simultaneously in the classification process ('four-view classification').

One-View Tumour Classification in Heterogeneous Breast Tissue

The SOM is trained using DWT coefficients corresponding the 276 tumours located in heterogeneous breast tissue. Some sample SOM images for this test data are shown in Fig. 5.32a, b.

   Due to the dielectric heterogeneity of the breast tissue, the classification process becomes much more difficult. The SOM no longer forms three distinct regions corresponding to the three different tumour types, as shown in Fig. 5.32a, b. Therefore the SOM can no longer be used for either tumour classification or tumour tracking. Therefore a more robust method for SOM classification/tracking must be developed. This motivates the investigation of the four-view classification approach.

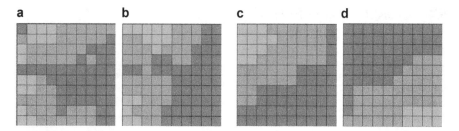

Fig. 5.32 Performance of the SOM tracking algorithm for tumours in a dielectrically heterogeneous breast. **(a)** and **(b)** correspond to one-view SOMs, while **(c)** and **(d)** correspond to four-view SOMs)

Four-View Tumour Classification in Heterogeneous Breast Tissue

In the four-view classification algorithm, the DWT coefficients corresponding to the backscattered signal received at each of the four antennas surrounding the tumour are combined to improve classification performance. Rather than using 15 DWT coefficients as input into the SOM, this algorithm uses 60 coefficients (4 × 15). The data set is thus quartered in size and the SOM is trained using the DWT coefficients corresponding to 69 tumours, and tested using 21 separate and distinct tumours. The average performance across ten shuffled files was as follows:

- 100 % for macrolobulated tumours
- 92.29 % for 3 spiculated malignant tumours
- 89.86 % for 10 spiculated malignant tumours

yielding an overall average accuracy of 94.05 %.

Using the four-view approach, the SOM forms three clear distinct regions, corresponding to the three different tumour types, as shown in Fig. 5.32c, d. This is reflected in the overall classification accuracy of 94.05 % across all tumour types. This result is particularly promising given the fact that the tumour was located very close to a region of dielectric heterogeneity.

In this section, 'Tumour Tracking Maps' have shown to have the ability to differentiate between macrolobulated benign and two different levels of malignant tumours. Therefore, these 'Tumour Tracking Maps' have significant potential as a cancer classification or diagnosis tool, even in dielectrically heterogeneous breasts.

However, more importantly they could also be used to monitor the development of a tumour due to the fact that 'Tumour Tracking Maps' preserve the topology of the input information. Therefore, a clinician could use these maps to determine whether a tumour is developing from benign to malignant (moving across the 'Tumour Tracking Maps') or not (staying static on the 'Tumour Tracking Map'). Movement even within the benign region could indicate that the tumour is developing even before it is classified as malignant and so treatment could be offered to patients at a very early stage of their disease, when it is most effective.

5.4 Classification of Early-Stage Breast Cancer
in 3-D Experimental Results

For this subsection, a pre-clinical UWB prototype imaging system from the
University of Manitoba with tumour and breast phantoms were used, and are
described in the following sections.

5.4.1 Experimental Setup and Materials

The system is described in detail elsewhere [21], but a short description will be
provided here for completeness. The system comprised a Vector Network Analyser
(VNA) which was connected to a computer via Ethernet. The system also comprised
an in-house developed Vivaldi antenna which was connected to the VNA via a 50 Ω
cable and was attached to a wall in the interior of a Plexiglas tank filled (with canola
oil). The VNA was a Field Fox N9923A model from Agilent Technologies. The
monostatic Vivaldi antenna was manufactured with two layers of Arlon–Diclad 527,
with a permittivity of 2.65 and a loss tangent of 0.0022. The antenna was mounted
into a waterproof acrylic mounting structure so that it could be immersed in the
Plexiglas tank filled with canola oil.

The antenna had a fixed position (attached to a wall of the Plexiglass tank)
throughout measurements but the phantoms were attached and rotated using a step-
motor via custom-made Pyrex thin pipes which have similar dielectric properties
to canola oil. When placing the breast phantom, the tumour phantom was always
located at the shortest distance possible from the antenna. The step-motor was set to
rotate at every 2.5°, and so allowing for 144 independent measurements around each
tested phantom. The radius of the scan geometry was 21 cm. A full measurement
around the phantom took approximately 120 s. Finally, it is worth mentioning that
the pre-clinical prototype was isolated from Wi-Fi and mobile network interference.

The Plexiglas tank was filled with a coupling medium canola oil which provided
a similar speed of propagation at microwave frequencies to that within the breast
phantom [71]. The breast phantom was simulated using a styrene-acrylonitrile
cylinder with a diameter of 13 cm and a height of 35 cm filled with glycerine,
since the dielectric permittivity of glycerine mimics the average values found in
low density breast regions [45]. Tumour phantoms were simulated with a mixture
of TX151 solidifying powder on a volume proportion of 6:1 of water to TX151
powder. Fibroglandular cluster phantoms, which were introduced to account for
the heterogeneity of the breast phantom, were simulated with a mixture of TX151
solidifying powder on a volume proportion of 4:1 of water to TX151 powder.

A total of 15 malignant and 20 benign tumour phantoms were modelled with
the above-mentioned mixture of TX151 and water. The shapes were modelled to
approximate the mathematical GRS phantoms previously mentioned in Sect. 5.1.2:
spiculated and microlobulated shapes represent malignant tumours, whereas round

Table 5.6 Dielectric properties of the breast tissue tumour phantom mimicking materials and coupling medium at 3 GHz

Material	Mimicked tissue	ε_r	σ (S/m)
Styrene-acrylonitrile	Skin	2.6	0.001
Canola oil	Coupling medium	2.5	0.035
Glycerine	Low density breast tissue	8.9	0.72
4:1 of water:TX151	Fibroglandular tissue	25–32	1.5
6:1 of water:TX151	Tumour	55	2.1

and oval shapes represent benign tumours. A sub-set of the data set with 13 benign and 13 malignant tumour phantoms with average diameters ranging between 13 and 40 mm was selected from these phantoms. Fibroglandular cluster phantoms were modelled with the respective mixture of TX151 and water: a rough cylinder shape with 1.5 cm of diameter and 3 cm of height. The tissue mimicking materials used in these phantoms represent the actual dielectric properties of the biological tissues in the frequency range between 1 and 6 GHz. Table 5.6 lists the dielectric properties for the different used materials [21] at 3 GHz.

5.4.2 Methodology

The methodology is divided in the following sections: data acquisition and data post-processing.

5.4.2.1 Data Acquisition

A monostatic radar system was used, in which a single antenna was used as a transceiver, and as a result the VNA functioned as both a microwave waveform generator and also as a recording device for the reflected waves. A frequency-sweep from 1 to 6 GHz was applied to each tumour phantom at 144 different angular positions, for the homogeneous and heterogeneous breast phantoms. In a clinical system, an antenna would rotate around the breast and record the tumour reflection at a range of angular positions. However, for the present system, the entire phantom was rotated while the antenna was fixed (144 rotations, of 2.5° each, are considered). A total of 26 different tumour phantoms were tested (13 malignant and 13 benign tumour models) in the two breast phantoms, the homogeneous and the heterogeneous, a total of 3744 backscattered signals for each breast phantom were recorded for use with the classification algorithm. It should be noted that the tumour models and fibroglandular cluster models were positioned at different locations within the breast phantom.

5.4.2.2   Data Post-Processing

Five steps will be completed towards the data post-processing.

Change from Frequency Domain to Time Domain

Since the VNA collected the reflections in the frequency domain (FD), an inverse fast Fourier transform (IFFT) was applied to all signals to convert them into the time domain (TD) prior to further processing. The RTS of the tumours was then extracted from the TD signals with a FEM.

Feature Extraction Methods

Once the signals were converted to time domain, PCA was used to change the backscattered data from the tumours into a new orthonormal basis with maximised variance, and so extracting the most significant components of the tumours backscatter, which defines the RTS of each tumour. The dimensionality reduction of PCA was investigated, so that unnecessary computational complexity could be avoided while maximising classification performance.

Classification Algorithms

The following classifiers were used in this study: LDA, QDA and a machine learning algorithm: SVM, which were fully described in the following Sects. 5.3.2.1, 5.3.2.2 and 5.3.2.3.

Classification Architecture

Since the presented tumour databases only included two classes of tumour phantoms (benign and malignant), the following simple classification architecture (Fig. 5.26 on the left) was used: Coarse-Shape (CS).

K-Fold Cross-Validation

Classifiers based on LDA, QDA and SVM can be applied to the data using the K-fold cross-validation method so that the performance of each classifier is evaluated while assuring that the training and testing groups for the classifier are treated independently, and so overfitting is avoided. The data was divided in $K = 13$ groups for cross-validation, each group comprised 144 signals from 2 different tumours (1 benign and 1 malignant). Reflected signals from each two tumours (total of

288 signals) were tested against the remaining 24 training tumours (total of 3456 signals), then the reflected signals from two other tumours were tested against the remaining 24 training tumours, and so on until all tumours were tested for classification.

5.4.3 Results

The results are sub-divided and presented in the following sections.

5.4.3.1 Principal Component Selection

Firstly, all signals are calibrated (to ensure that background and electronic noise is reduced to a minimum) and are transformed from FD to TD using an IFFT.

Then, the dimensionality reduction, through PCA, is investigated. The classification performance of tumour phantoms was obtained for varying numbers of principal components, using both LDA and QDA classifiers, for both breast phantom (homogeneous and heterogeneous) scenarios, without cross-validation. A subset of 13 principal components is selected so that unnecessary computational complexity can be avoided while maintaining high classification performance. The number of 13 principal components will be used for the remainder of the chapter.

5.4.3.2 Antenna Selection

As above mentioned, the backscattered signals are recorded at 144 different angular positions. The first antenna location ($0°$) is the antenna located closest to the tumour phantom. The backscattered signals recorded at this antenna location are expected to have more information on the RTS of the tumour model and less of the breast phantom and noise from the system and coupling medium. This is mainly due to a shorter propagation path between the antenna and tumour model. The performance of the classification taking only into account the signal recorded at the 1st location ($0°$) was compared to that taking only into account the backscattered signal recorded at the 73rd position ($180°$)—results presented in Table 5.7.

While the 1st antenna often yielded 100 % classification performance in all breast phantoms and with both LDA and QDA classifiers, the full 144 backscattered signals will be used in the remainder of this section. As a result, it is assumed for the remainder of the chapter that the location of the tumour is not previously known, and that more noise (yielding from longer propagation paths) is introduced into the classification algorithms.

5.4.3.3 Data Post-Processing

The performance for the SVM-based classification architectures depends on optimal values for parameters γ and C, as described in Sect. 5.3.2.3. The selection of these parameters was completed using an empirical grid-search, and γ and C were tested between the range of 2^{-15} and 2^{15}. For the results presented here, the optimal SVM parameters were determined at $\gamma = 2^2$ and $C = 2^3$, for both the tumour databases in the homogeneous and the heterogeneous breast phantoms.

The classification performance obtained with LDA, QDA and SVM with a Coarse-Shape classification architecture (to classify benign and malignant tumours), is shown in Table 5.8, for the tumour databases in the homogeneous and heterogeneous breast phantoms.

In this study, the recorded signals were treated independently and also in a combined way. In the first and third rows of Table 5.8, the recorded signals are treated independently, i.e. there is no record of which of the 144 recorded backscattered signals belong to the 26 tumours, and so each reflection is treated as a separate test to the classifier.

The classification results obtained by combining the backscattered signals for each tumour are shown in the second and fourth rows of Table 5.8, for the homogeneous and heterogeneous breast phantoms, respectively. For combined classification, the backscattered signals recorded at 144 angular positions around the breast for each tumour are classified, and the classification of this tumour is a simple majority of all results for all antennas.

It is observed that, even though the 26 tumour phantoms considered here have varying sizes, the CS classification architecture yields high classification performance, up to 90.95 and 96.15 % with SVM, when not combining and when combining the 144 signals for each tumour, respectively, for the homogeneous breast phantom. SVM outperformed LDA by 3.85 % and QDA by 1.61 % when classified signals were considered independently. SVM outperformed LDA by 7.69 % and QDA by 3.84 % when classified signals were combined.

Even though the classification of tumours in the heterogeneous breast phantom is more challenging, the classification still reaches 87.07 and 92.31 % when not combining and when combining the 144 signals for the classification of each

Table 5.7 Classification accuracies (%) with the closest and furthest antenna to the tumour, in both homogeneous and heterogeneous breast phantoms

Antenna	Homogeneous		Heterogeneous	
	LDA	QDA	LDA	QDA
1st at 0°	100	100	100	100
73rd at180°				
	86.8	98.9	91.5	99.9

Courtesy of Hugo Medeiros and Faculdade de Ciências e Tecnologia, Universidade Nova de Lisboa [46]

tumour, respectively. SVM outperformed LDA by 2.78 % and QDA by 0.48 % when classified signals were considered independently. SVM outperformed LDA by 7.69 % and QDA by 3.85 % when classified signals were combined.

5.4.4 Conclusions

This chapter considered the classification of benign and malignant tumours, based solely on their RTS, in both homogeneous and heterogeneous breast phantoms. Results presented here are in general agreement with previous simulated data in Sect. 5.3.4, and clearly show the need for further experimental measurements. Tumour phantoms within the breast phantoms were subjected to a coarse-shape classifier and their classification performance was as high as 96.15 % when using SVM.

In terms of global classification performance, LDA was consistently outperformed by QDA, which, in turn, was outperformed by SVM. This was observed in both classification scenarios (for tumours embedded in both homogeneous and heterogeneous breast phantoms), and also when considering independent and combined signals.

5.5 Emerging Approaches: Classification of Differential Signals to Monitor Changes within the Breast

The authors would like to acknowledge the work completed at McGill University, Canada [37, 62–64] regarding the use of classification algorithms to monitor changes within the breast. Researchers in this group have successfully shown a number of studies using both LDA and SVM classification algorithms used to this end. In particular in [37], a complex cost-sensitive ensemble classifier was presented, consisting of three main components: feature extraction, classification

Table 5.8 Classification accuracies (%) for a coarse-shape classification architecture using linear discriminant analysis (LDA), quadratic discriminant analysis (QDA) and support vector machines (SVM) in homogeneous and heterogeneous breast phantoms

Phantom	Signals	LDA	QDA	SVM
Homogeneous	Independent	87.1	89.3	90.9
	Combined	88	92	96
Heterogeneous	Independent	84.3	86.6	87.1
	Combined	85	88	92

Courtesy of Hugo Medeiros and Faculdade de Ciências e Tecnologia, Universidade Nova de Lisboa [46]

and fusion. The aim of this chapter was to provide classification on detected lesions, and so for more information on these studies, the interested reader is referred to the above-mentioned studies.

References

[1] Bartholomew DJ, Steele F, Moustaki I, Galbraith JI (2002) The analysis and interpretation of multivariate data for social scientists. CRC Press, Florida, USA
[2] Bennett KP, Campbell C (2000) Support vector machines: hype or hallelujah? ACM SIGKDD Explorations Newsletter 2(2):1–13
[3] Bond EJ, Li X, Hagness SC, Van Veen BD (2003) Microwave imaging via space-time beam-forming for early detection of breast cancer. Antennas and Propagation, IEEE Transactions on 51(8):1690–1705
[4] Boser BE, Guyon IM, Vapnik VN (1992) A training algorithm for optimal margin classifiers. In: Proceedings of the Fifth Annual Workshop on Computational Learning Theory, ACM, pp 144–152
[5] Campbell C (2008) Introduction to support vector machines. http://videolectures.net/epsrcws08_campbell_isvm/
[6] Chambers D, Berryman J (2006) Target characterization using decomposition of the time-reversal operator: electromagnetic scattering from small ellipsoids. Inverse Problems 22(6):2145–2163
[7] Chen CC, Peters Jr L (1997) Buried unexploded ordnance identification via complex natural resonances. Antennas and Propagation, IEEE Transactions on 45(11):1645–1654
[8] Chen Y, Gunawan E, Low KS, Wang SC, Soh CB, Putti TC (2008) Effect of lesion morphology on microwave signature in 2-d ultra-wideband breast imaging. Biomedical Engineering, IEEE Transactions on 55(8):2011–2021
[9] Chen Y, Craddock IJ, Kosmas P (2010a) Feasibility study of lesion classification via contrast-agent-aided UWB breast imaging. Biomedical Engineering, IEEE Transactions on 57(5):1003–1007
[10] Chen Y, Craddock IJ, Kosmas P, Ghavami M, Rapajic P (2010b) Multiple-input multiple-output radar for lesion classification in ultrawideband breast imaging. Selected Topics in Signal Processing, IEEE Journal of 4(1):187–201
[11] Conceição R (2010) The development of ultra wideband scanning techniques for detection and classification of breast cancer. PhD thesis, National University of Ireland, Galway
[12] Conceição R, O'Halloran M, Glavin M, Jones E (2011a) Evaluation of features and classifiers for classification of early-stage breast cancer. Journal of Electromagnetic Waves and Applications 25(1):1–14
[13] Conceição RC, O'Halloran M, Glavin M, Jones E (2010) Support vector machines for the classification of early-stage breast cancer based on radar target signatures. Progress In Electromagnetics Research B 23:311–327, DOI 10.2528/PIERB10062407, http://www.jpier.org/pierb/pier.php?paper=10062407
[14] Conceição RC, O'Halloran M, Glavin M, Jones E (2011b) Effects of dielectric heterogeneity in the performance of breast tumour classifiers. Progress In Electromagnetics Research M 17:73–86, DOI 10.2528/PIERM10122402, http://www.jpier.org/pierm/pier.php?paper=10122402
[15] Conceição RC, O'Halloran M, Glavin M, Jones E (2011c) Numerical modelling for ultra wideband radar breast cancer detection and classification. Progress In Electromagnetics Research B 34:145–171, DOI 10.2528/PIERB11072705, http://www.jpier.org/pierb/pier.php?paper=11072705
[16] Cortes C, Vapnik V (1995) Support-vector networks. Machine learning 20(3):273–297

[17] Davis SK, Van Veen BD, Hagness SC, Kelcz F (2008) Breast tumor characterization based on ultrawideband microwave backscatter. Biomedical Engineering, IEEE Transactions on 55(1):237–246

[18] Deboeck G, Kohonen T (1998) Visual explorations in finance: with self-organizing maps. Springer-Verlag London Ltd., UK

[19] El-Shenawee M, Miller EL (2006) Spherical harmonics microwave algorithm for shape and location reconstruction of breast cancer tumor. Medical Imaging, IEEE Transactions on 25(10):1258–1271

[20] Everitt BS, Dunn G (2001) Applied multivariate data analysis, 2nd edn. Arnold London, UK

[21] Flores-Tapia D, Pistorius S (2011) Real time breast microwave radar image reconstruction using circular holography: A study of experimental feasibility. Medical physics 38(10):5420–5431

[22] Gävert H, Hurri J, Särelä J, Hyvärinen A (2010) The fastica package for matlab. http://www.cis.hut.fi/projects/ica/fastica

[23] Ghavami M, Michael LB, Haruyama S, Kohno R (2002) A novel uwb pulse shape modulation system. Wireless Personal Communications 23(1):105–120

[24] Goldberg DE, Richardson J (1987) Genetic algorithms with sharing for multimodal function optimization. In: Genetic algorithms and their applications: Proceedings of the Second International Conference on Genetic Algorithms, pp 41–49

[25] Haimovich AM, Blum RS, Cimini LJ (2008) MIMO radar with widely separated antennas. Signal Processing Magazine, IEEE 25(1):116–129

[26] Hair JF, Black WC, Babin BJ, Anderson RE, Tatham RL (2006) Multivariate data analysis, 6th edn. Pearson Prentice Hall Upper Saddle River, New Jersey, USA

[27] Holland JH (1992) Adaptation in natural and artificial systems: an introductory analysis with applications to biology, control, and artificial intelligence. MIT Press, Cambridge, MA, USA

[28] Hsu CW, Chang CC, Lin CJ (2003) A practical guide to support vector classification. National Taiwan U. http://www.csie.ntu.edu.tw/cjlin/papers/guide/guidepdf

[29] Hyvärinen A, Oja E (2000) Independent component analysis: algorithms and applications. Neural networks 13(4–5):411–430

[30] Jones M, Byrne D, McGinley B, Morgan F, Glavin M, Jones E, O'Halloran M, Conceição RC (2013) Classification and monitoring of early stage breast cancer using ultra wide band radar. In: The Eighth International Conference on Systems (ICONS), International Academy, Research, and Industry Association (IARIA), pp 46–51

[31] Kohonen T (1990) The self-organizing map. Proceedings of the IEEE 78(9):1464–1480

[32] Kosmas P, Rappaport CM (2006) FDTD-based time reversal for microwave breast cancer detection-localization in three dimensions. Microwave Theory and Techniques, IEEE Transactions on 54(4):1921–1927

[33] Kosmas P, Laranjeira S, Dixon J, Li X, Chen Y (2010) Time reversal microwave breast imaging for contrast-enhanced tumor classification. In: Engineering in Medicine and Biology Society (EMBC), 2010 Annual International Conference of the IEEE, pp 708–711

[34] Krzanowski WJ (1988) Principles of Multivariate Analysis: A User's Perspective. Oxford Statistical Science Series. Oxford University Press, New York, USA

[35] Lazebnik M, Popovic D, McCartney L, Watkins CB, Lindstrom MJ, Harter J, Sewall S, Ogilvie T, Magliocco A, Breslin TM, et al (2007) A large-scale study of the ultrawideband microwave dielectric properties of normal, benign and malignant breast tissues obtained from cancer surgeries. Physics in Medicine and Biology 52(20):6093–6115

[36] Li J, Stoica P (2007) Mimo radar with colocated antennas. Signal Processing Magazine, IEEE 24(5):106–114

[37] Li Y, Santorelli A, Laforest O, Coates M (2015) Cost-sensitive ensemble classifiers for microwave breast cancer detection. In: 2015 IEEE International Conference on Acoustics, Speech and Signal Processing (ICASSP), IEEE, pp 952–956

[38] Maass W (1997) Networks of spiking neurons: the third generation of neural network models. Neural networks 10(9):1659–1671

[39] Maass W (1999) Computing with spiking neurons. Pulsed Neural Networks

[40] Mallat SG (1989) A theory for multiresolution signal decomposition: the wavelet representation. Pattern Analysis and Machine Intelligence, IEEE Transactions on 11(7):674–693

[41] Mallet Y, Coomans D, Kautsky J, De Vel O (1997) Classification using adaptive wavelets for feature extraction. Pattern Analysis and Machine Intelligence, IEEE Transactions on 19(10):1058–1066

[42] Mashal A, Booske JH, Hagness SC (2009a) Toward contrast-enhanced microwave-induced thermoacoustic imaging of breast cancer: An experimental study of the effects of microbubbles on simple thermoacoustic targets. Physics in medicine and biology 54(3):641–650

[43] Mashal A, Sitharaman B, Booske JH, Hagness SC (2009b) Dielectric characterization of carbon nanotube contrast agents for microwave breast cancer detection. In: Antennas and Propagation Society International Symposium, 2009. APSURSI'09. IEEE, pp 1–4

[44] McGinley B, O'Halloran M, Conceição RC, Morgan F, Glavin M, Jones E (2010) Spiking neural networks for breast cancer classification using radar target signatures. Progress In Electromagnetics Research C 17:79–94

[45] Meaney PM, Fanning MW, Raynolds T, Fox CJ, Fang Q, Kogel CA, Poplack SP, Paulsen KD (2007) Initial clinical experience with microwave breast imaging in women with normal mammography. Academic Radiology 14(2):207–218

[46] Medeiros HF (2013) Classificação de tumores de cancro na mama através de radar de banda ultra-larga de microondas. PhD thesis, Faculdade de Ciências e Tecnologia, Universidade Nova de Lisboa, http://run.unl.pt/bitstream/10362/11483/1/Medeiros_2013.pdf

[47] Mishchenko MI, Hovenier JW, Travis LD (1999) Light scattering by nonspherical particles: theory, measurements, and applications, Academic press, chap Chapter 11: Light scattering by stochastically shaped particles

[48] Muinonen K (1998) Introducing the gaussian shape hypothesis for asteroids and comets. Astronomy and Astrophysics 332:1087–1098

[49] Ng A (2010) Support vector machines (part v of cs229 machine learning course materials). http://www.stanford.edu/class/cs229/notes/cs229-notes3.pdf

[50] O'Halloran M, Cawley S, McGinley B, Conceição RC, Morgan F, Jones E, Glavin M (2011a) Evolving spiking neural network topologies for breast cancer classification in a dielectrically heterogeneous breast. Progress In Electromagnetics Research Letters 25:153–162, DOI 10.2528/PIERL11050605, http://www.jpier.org/pierl/pier.php?paper=11050605

[51] O'Halloran M, McGinley B, Conceição RC, Morgan F, Jones E, Glavin M (2011b) Spiking neural networks for breast cancer classification in a dielectrically heterogeneous breast. Progress In Electromagnetics Research 113:413–428, DOI 10.2528/PIER10122203, http://www.jpier.org/pier/pier.php?paper=10122203

[52] Oliveira B, Glavin M, Jones E, O'Halloran M, Conceição R (2014) Avoiding unnecessary breast biopsies: Clinically-informed 3d breast tumour models for microwave imaging applications. In: Antennas and Propagation Society International Symposium (APSURSI), 2014 IEEE, IEEE, pp 1143–1144

[53] Oliveira B, O'Halloran M, Conceição R, Glavin M, Jones E (2015) Development of clinically-informed 3d tumor models for microwave imaging applications. accepted in IEEE Antennas and Wireless Propagation Letters DOI 10.1109/LAWP.2015.2456051

[54] Pande S, Morgan F, Cawley S, McGinely B, Carrillo S, Harkin J, McDaid L (2010) EMBRACE-SysC for analysis of NoC-based spiking neural network architectures. In: International Symposium on System-on-Chip, IEEE

[55] Prada C (2002) Detection and imaging in complex media with the D.O.R.T method. In: Imaging of complex media with acoustic and seismic waves, Springer, pp 107–134

[56] Rangayyan RM, El-Faramawy NM, Desautels JEL, Alim O, et al (1997) Measures of acutance and shape for classification of breast tumors. Medical Imaging, IEEE Transactions on 16(6):799–810

[57] Raykov T, Marcoulides GA (2008) An introduction to applied multivariate analysis. Routledge Taylor & Francis Group, New York, USA

[58] Richmond JH (1965) Scattering by a dielectric cylinder of arbitrary cross section shape. Antennas and Propagation, IEEE Transactions on 13(3):334–341

[59] Rocke P, McGinley B, Maher J, Morgan F, Harkin J (2008) Investigating the suitability of FPAAs for evolved hardware spiking neural networks. In: Evolvable Systems: From Biology to Hardware, Springer, pp 118–129

[60] Roussinov DG, Chen H (1998) A scalable self-organizing map algorithm for textual classification: A neural network approach to thesaurus generation. Communication and Cognition in Artificial Intelligence Journal 15(1–2):81–112

[61] Saito N, Coifman RR (1995) Local discriminant bases and their applications. Journal of Mathematical Imaging and Vision 5(4):337–358

[62] Santorelli A, Li Y, Porter E, Popović M, Coates M (2014a) Image classification for a time-domain microwave radar system: Experiments with stable modular breast phantoms. In: 2014 8th European Conference on Antennas and Propagation (EuCAP), IEEE, pp 320–324

[63] Santorelli A, Porter E, Kirshin E, Liu YJ, Popović M (2014b) Investigation of classifiers for tumor detection with an experimental time-domain breast screening system. Progress In Electromagnetics Research 144(2):45–57

[64] Santorelli A, Laforest O, Porter E, Popović M (2015) Image classification for a time-domain microwave radar system: Experiments with stable modular breast phantoms. In: 2015 9th European Conference on Antennas and Propagation (EuCAP), IEEE, pp 1–5

[65] Saunders R, Samei E, Baker J, Delong D (2006) Simulation of mammographic lesions. Academic radiology 13(7):860–870

[66] Scott MJJ, Niranjan M, Prager RW (1998) Realisable classifiers: Improving operating performance on variable cost problems. In: BMVC, pp 306–315

[67] Seber GAF (1984) Multivariate observations. John Wiley & Sons, New Jersey, USA

[68] Serrano-Cinca C (1996) Self organizing neural networks for financial diagnosis. Decision Support Systems 23(3):227–238

[69] Shea JD, Kosmas P, Hagness SC, Van Veen BD (2009) Contrast-enhanced microwave breast imaging. In: Antenna Technology and Applied Electromagnetics and the Canadian Radio Science Meeting, 2009. ANTEM/URSI 2009. 13th International Symposium on, IEEE, pp 1–4

[70] Shlens J (2003) A tutorial on principal component analysis. http://www.cs.princeton.edu/picasso/mats/PCA-Tutorial-Intuition_jp.pdf

[71] Sill JM, Fear EC (2005) Tissue sensing adaptive radar for breast cancer detection-experimental investigation of simple tumor models. Microwave Theory and Techniques, IEEE Transactions on 53(11):3312–3319

[72] Stanley KO, Miikkulainen R (2002) Evolving neural networks through augmenting topologies. Evolutionary computation 10(2):99–127

[73] Teo J, Chen Y, Soh CB, Gunawan E, Low KS, Putti TC, Wang SC (2010) Breast lesion classification using ultrawideband early time breast lesion response. Antennas and Propagation, IEEE Transactions on 58(8):2604–2613

[74] Valens C (1999–2010) A really friendly guide to wavelets. http://pagesperso-orange.fr/polyvalens/clemens/wavelets/wavelets.html#section1

[75] Veeravalli VV, Basar T, Poor HV (1994) Minimax robust decentralized detection. Information Theory, IEEE Transactions on 40(1):35–40

[76] Wasilewski F (2010) Pywavelets - coiflets 5 wavelet (coif5) properties, filters and functions - wavelet properties browser. http://wavelets.pybytes.com/wavelet/coif5

[77] Wickerhauser MV (1994) Adapted wavelet analysis from theory to software. AK Peters Ltd., MA, USA

[78] Wold H, et al (1966) Multivariate analysis, Academic Press, New York, USA, chap Estimation of principal components and related models by iterative least squares, pp 391–420

Chapter 6
Experimental Systems

Johan Jacob Mohr and Tonny Rubæk

In this chapter, prototype microwave breast imaging systems will be presented. Rather than focusing on historical developments, preference will be given to systems which are now fully operational and have been evaluated using patients. At time of printing, only the prototype systems at Dartmouth College, the University of Bristol, the University of Calgary, McGill University, and the Electronics and Telecommunications Research Institute (ETRI), Korea, have reached an operational level. These systems are described first in Sect. 6.1. The prototype animal imaging system from Carolinas Medical Center is also included since it exhibits many similarities with breast cancer systems. Note: the prototype system from the Technical University of Istanbul and the company Mitos has also been trialled on patients, but full details of the system and corresponding clinical results have not been published and therefore it has not been included in this chapter.

For completeness, a representative range of other prototype systems currently in development are briefly described in subsequent sections. These systems are divided into tomographic systems, described in Sect. 6.2, while radar-based systems are described in Sect. 6.3.

J.J. Mohr (✉)
formerly at the Department of Electrical Engineering, Technical University of Denmark, Kongens Lyngby, Denmark

Mellanox Technologies, Roskilde, Denmark
e-mail: johan_jacob_mohr@yahoo.dh

T. Rubæk
formerly at the Department of Electrical Engineering, Technical University of Denmark, Kongens Lyngby, Denmark

OHB Systems, Bremen, Germany
e-mail: trubaek@gmail.com

© Springer International Publishing Switzerland 2016   131
R.C. Conceição et al. (eds.), *An Introduction to Microwave Imaging for Breast Cancer Detection*, Biological and Medical Physics, Biomedical Engineering, DOI 10.1007/978-3-319-27866-7_6

A system that does not fit into these categories is one from the Middle East Technical University, Ankara, Turkey. In their method denoted harmonic motion microwave doppler imaging a focused ultrasound probe mechanically modulates the tissue, which in turn phase modulates the backscattered and received microwave signals, [71]. Results were presented in [68] and at MiMed COST meetings (TD1301).

6.1 Operational Systems

Six systems with the characteristics summarised in Table 6.1 are described here. The tomographic system for breast cancer detection at Dartmouth College has been used for hundreds of patients and has shown good clinical results. The University of Bristol system uses a very different ultra-wideband radar technique and reports from clinical experiments on 95 patients are promising. The radar-based systems from the University of Calgary and the McGill University have been used for clinical experiments on 9 patients and for 13 healthy patients, respectively. Recent reports on the system from the Electronics and Telecommunications Research Institute (ETRI), Korea, indicate that this system is patient-ready and it has been used for animal experiments. Finally, different configurations of the Carolinas Medical Center system, used for animal experiments, are described.

The system at Dartmouth College results from more than a decade of development, with the first clinical test reported as early as the year 2000 [43, 44, 53].

Table 6.1 Selected operational microwave imaging systems

Origin	Type	Key features	Results
Dartmouth College	Tomographic	2 × 8 antennas mechanical scanning laser/camera system	500+ patients
University of Bristol	UWB radar	60 antennas ceramic fitting cups 10 s acquisition time	95 patients (with this system)
University of Calgary	Cylindrical tank (w. coupling liquid)	1 antenna mechanical scanning laser for outline	Pilot clinical experiments 9 patients
McGill University	Hemispherical ceramic cup/radome	16 antennas two perpendicular arcs	13 healthy patients (extended period)
ETRI, Korea	Tomographic	16 antennas Dartmouth inspired	Dogs Suitable for humans
Carolinas Medical Centre[a]	Tomographic	24 antennas 2-D dynamic phenomena	Anesthetised pig (foreleg)

[a] Another larger tank, used for euthanised animals, with mechanically scanned antennas allows for 3-D imaging

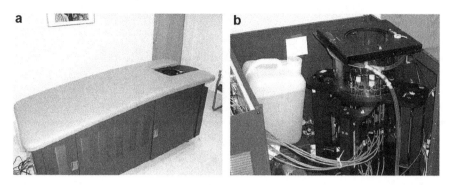

Fig. 6.1 (**a**) Dartmouth college system as presented to the patient. (**b**) The tank, antenna and pumping system. (Courtesy of Paul Meaney, Dartmouth College)

The latest system, see Fig. 6.1, uses 16 vertical monopole antennas, in a tank with a glycerine/water coupling liquid [14, 23]. The antennas form two interleaved sets, which can be positioned individually vertically. For each antenna position and frequency, the 240 data-points—16 Transmit Antennas (Tx) and 15 Receive Antennas (Rx)—can be acquired in about 6 s, using dedicated hardware. The hardware and antennas allow for frequencies up to 3 GHz, but best results are reported at 1300 MHz. This third generation system is modular in design, which also allows for connection to a more compact antenna system aiming at combining microwave tomography with magnetic resonance imaging (MRI) [45]. Two lasers and one camera on a circular gantry can record the shape of the breast [52]. This information is used during image formation to increase image quality, since the imaging zone in that case can be reduced to the volume within the known boundary of the breast [42]. In a later version of the inversion reconstruction algorithms the discrete dipole approximation (DDA) is used [23]. This allows for a much faster calculation of the electrical field as compared to the previously used FDTD approach so that inversion times become 5–10 s in the 2D case and roughly 10 min for a 3D image. Over 500 patient exams have been conducted; some aiming at detecting malignant tumours others aiming at monitoring chemotherapy [46]. An example image is shown in Fig. 6.2.

The radar-based University of Bristol system is also the latest generation in an evolution, starting with 16 antennas initially, to an intermediate system with 31 antennas, to their current system with 60 antennas [30, 35, 36]. The cavity-backed UWB slot antennas are mounted in a hemispherical acrylonitrile butadiene styrene plastic half-sphere, which fits the breast via a low-loss ceramic fitting cup. A paraffin-based coupling medium is used to eliminate air gaps both between the antennas and the fitting cup and between the fitting cup and the breast. An 8-port, 8 GHz, vector network analyser (VNA) feeds the antennas through a 60-way, custom made, electromechanical switch, giving an acquisition time of 10 s (1770 signal paths). A delay-and-sum radar beamformer algorithm is used for image formation. Initial results from measurements on about 95 symptomatic patients are reported in [30].

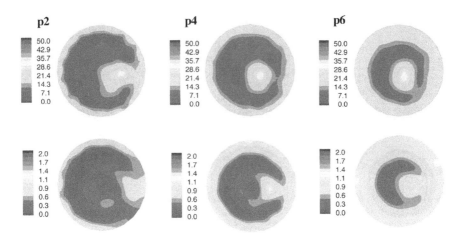

Fig. 6.2 Reconstruction of permittivity (*top*) and conductivity (*bottom*) from measurements on the left breast of one patient. The results are from planes 2, 4 and 6 (of 7) at 1300 MHz. The breast outline is seen to decrease as the distance from the chest increases. Also, it appears that the tumour is showing in both the permittivity and conductivity images; very often the permittivity is just indicating fibroglandular tissue. (Courtesy of Paul Meaney, Dartmouth College)

The system at University of Calgary is mechanically and electrically suitable for patient measurements. Pilot clinical experiments have been carried out [11, 13] and recently results from 8 patients have been reported [15]. The monostatic system with mechanical scanning uses one balanced antipodal Vivaldi antenna [9] in a tank with canola oil coupling liquid, combined with a laser to record the breast outline [73]. Data acquired with a VNA at frequencies between 50 MHz and 15 GHz are processed using a delay-and-sum beamformer. In addition to the measurements described in [11, 13], a dedicated two-antenna through-the-breast transmission measurements system aimed at determining bulk electrical parameters are reported in [10]. The results from patients measurements were mixed, but several of the image results were consistent with observations from mammograms and MRI scans [15].

The system at McGill University uses 16 antennas embedded in a ceramic (Al_2O_3) hemispherical radome with an inner radius of 7 cm, [60]. A 70 ps (FWHM) Gaussian pulse is shaped with a synthesised broadband reflector before amplification resulting in a 2–4 GHz signal. An oscilloscope records the received time-domain data. The system has been tested, [55], on phantoms of realistic shape and heterogeneous content of tissue mimicking fat, skin, gland, and tumour [54]. The antenna system is small enough to allow for an easy integration into a portable examination bench [56]. The system has now been used repeatedly on 13 healthy patients over an extended 8-month period, e.g. [57, 58], to assess patient and measurement variability in a clinical setting.

The tomographic Electronics and Telecommunications Research Institute (ETRI) system, Korea, is inspired by the one at Dartmouth College. It has 16 vertical monopole antennas in a plane circular arrangement [37, 65]. The antennas can be moved vertically in a cylindrical tank with a propylene glycol coupling liquid. The operation frequency range is 0.5–3 GHz. A continuous wave (CW) signal from a signal source is distributed and received through switching networks and the down-converted Intermediate Frequency (IF) signals and reference are sampled by a multi-channel analog-to-digital (A/D) converter. Image formation from each acquisition plane is completed in 2-D by using an FDTD forward solver in conjunction with a regularised Gauss–Newton inversion algorithm. Performance has been evaluated on cylindrical and spherical phantoms [37, 65] with 5 mm radius cylindrical and 5 mm radius spherical water inclusions. As a next step towards clinical trials, 5 dogs where examined both with the microwave tomographic system and MRI [41]. Two had no existing breast tumours, one had a tumour, and two had suspected tumours. Results aligned with MRI findings, but the estimated tumour sizes differed.

The main focus of the microwave imaging systems at Carolinas Medical Center has been measurements on animals. One version uses a tank with a 120 cm diameter and 135 cm height. 2-D and 3-D tomographic images of a dead dog are reported in [63]. The mechanical scanning of two antennas, though, resulted in very long acquisition times (9 h), and an upgraded system using 1 transmitting antenna and a 16-receiving-antenna array was designed [64]. The antennas were designed for 0.8–1.2 GHz and a VNA used for data acquisition. 2-D and 3-D tomographic images of a swine torso obtained with this system at 0.9 GHz are provided in [64]. Another 2-D configuration for detection of physiological activity and soft tissue intervention has also been developed [61]. It uses 24 ceramic filled wave guide antennas, placed along a horizontal cross section of a metallic tank with a 21.5 cm diameter. This antenna system allows for operation in the range 1.0–2.3 GHz. Electronic scanning can acquire the required 24×23 complex measurements in just 12 ms opening the possibility for detection of dynamic physiological phenomena, for example, variation of blood content. Results using an iterative Newton algorithm, as well as a Born method on measurements on the foreleg of an anesthetised swine, are reported in [62].

6.2 Tomographic Systems

The tomographic systems described are summarised in Table 6.2. Measurements on non-biological phantoms have been reported from all systems. The reason for the lack of reports of measurements on ex vivo biological targets (which have been conducted with at least three of the systems) is probably due to the fact that tissue and tumours are known to change electrical properties when excised and the fact that the properties of the inherently inhomogeneous biological target are difficult to measure by other means. Each of the selected systems and their results are briefly summarised. Further work is also worth mentioning, e.g. the Clemson University system which has been used for microwave imaging [12] and for clinical

Table 6.2 Selected tomographic microwave imaging systems

Origin	Imaging domain	Antenna configuration	Targets
Technical Univ. of Denmark	Cylindrical tank (w. coupling liquid)	32 monopoles 3-D arrangement	Water-filled spheres Misc. short cylinders
Chalmers Univ. of Technology	Cylindrical tank (w. coupling liquid)	32 monopoles 3-D arrangement	Short plastic cylinder in saline water
University of Manitoba (2-D)	Cylindrical tank (for coupling liquid)	24 Vivaldi antennas Circular array	2-D objects (rods) Human arms and bovine legs
University of Manitoba (3-D)	Cylindrical chamber Air (prep. for liquids)	12 dual pol. antennas 120 MST probes	PEC calibration targets Misc. nylon and wood
McMaster University[a]	Between two sheets (X-ray compatible)	2 horn antennas 2-D planar scanning	Alginate and glycerine spheres in glycerine-based phantom
Duke University	Rectangular tub (w. coupling liquid)	2 dipoles 2-D planar scanning	Clay spheres in water
University of Michigan	12-panel cavity (w. coupling liquid)	36 bow-ties 3 circular arrays	Acrylic spheres Also hyperthermia
Supélec[b]	Rectangular chamber (w. coupling liquid)	2 large horns w. retinas Target can be rotated	Tube w. water in cylinder w. Triton X-100 mixture

[a]Rather a holographic imaging algorithm
[b]Also operated in a quasi real-time mode using a spectral reconstruction algorithm

patient measurements [33], and the antenna work at the University of Wisconsin–Madison [1, 2] aimed at designing a system where the microwave scan can take place with the patient on the MRI table, just before or after an MRI examination. A system was also developed at the University of Technology, Sydney, Australia, but results are not widely available [74].

The Technical University of Denmark system uses an acrylic tank with a diameter of 40 cm and a glycerine/water coupling liquid [80]. Thirty-two monopole antennas are arranged in a circular array with 4 layers and a diameter of 16 cm. Two synchronised signal generators are used along with custom-built transceivers and a switching network. It is designed for 0.5–4 GHz, but currently loss limits the upper frequency to about 2 GHz. Targets have primarily been water-filled spheres of diameters ranging between 20 and 40 mm and short cylinders, e.g. [31, 32]. A method-of-moment forward solver, which additionally calculates the Jacobian matrix, is used in conjunction with a Newton minimisation algorithm [59]. Currently, a modified system mechanically, electrically and operationally suitable for patient measurements is being developed.

The Chalmers University of Technology system represents a development from a 2-D system with 20 monopoles, in a circular configuration [17, 51, 78], to a 3-D system using a circular tank, with 32 monopoles in a cylindrical configuration [16]. The time-domain ultra-wideband tomographic system uses an impulse generator with a full-width half-maximum (FWHM) duration of about 75 ps, combined with a digitising oscilloscope and a switching matrix [78]. The imaging algorithm seeks to minimise the norm of the difference between predicted and measured time-domain electrical field-strength, and is summarised in [16], where reconstructions from a dielectric cylinder in saline water are also presented.

The 2-D system from the University of Manitoba system uses 24 Vivaldi antennas (3–10 GHz) arranged in a circular array [20]. A VNA combined with a mechanical 2×24 cross-bar is used for data acquisition. An enhanced distorted born iterative method or, alternatively, the multiplicative regularised contrast source inversion method is used for image reconstruction. Additionally, different multi-frequency schemes are used [20]. Measurements on a wide range of dielectric and conducting cylinders as well as a special E-phantom have been conducted and images have been reconstructed in the frequency range 3–6 GHz [20–22]. Also, the acquisition hardware has been used for measurements at 0.8–1.2 GHz on human arms (living tissue) and bovine legs (ex vivo tissue) with 24 dipoles in a metallic tank with a salt-water coupling liquid [49]. To improve calibration and to potentially increase the number of independent measurements, experiments with the scattering probe technique have been initiated [48].

The new dual polarised 3-D system from the University of Manitoba system is based on the experiences from the 2-D system and utilise the modulated scattering technique (MST) [48]. The 12 transmit/receive antennas, each have one probe, are arranged in a plane circular layout with a radius of 11.5 cm, while upper and lower layers with a 9.75 cm radius each have 24 probe pairs, amounting to a total of 120 MST probes [5]. Data acquisition in one frequency, in the range 3–5 GHz can be completed in less than 1 min, but typically a series of frequencies is used [5, 50]. Scattering from a known object is used to calibrate measured data before inversion with a parallelised multiplicative regularised finite-element contrast source inversion algorithm [77]. Experimental imaging results from nylon cylinder and wooden blocks within a radius of 10 cm imaging domain with an air-background are presented in [5].

The system from McMaster University uses 2-D planar raster scanning, where two antennas (one Tx and one Rx) are moved together along the opposite sides of the imaging domain delimited by two rigid plates, giving a geometry similar to the one of X-ray mammography [3]. The TEM horn antennas are usable in the entire (UWB) band 3.1–10.6 GHz and a VNA is used for data acquisition. The 2-D and 3-D holographic imaging methods are reviewed in [4] and the de-blurring technique used to correct for non-point-wise antennas is reviewed in [3]. Measurements at 5, 7 and 9 GHz and 2-D reconstructions from a flat glycerine-based phantom with two embedded tumour-phantoms are presented in [3], and 7 GHz measurements and 2-D reconstructions from two spheres made of alginate powder embedded in a glycerine-based phantom are shown in [4].

The Duke University system uses mechanical planar scanning of two dipoles in a rectangular tub with coupling liquid [75]. Data are acquired by using a VNA. In the two-step inversion algorithm, an initial reconstruction is first calculated by combining the digital tensor approximation and the Born iterative method (BIM). Secondly, this is refined using the stabilised bi-conjugate gradient fast Fourier transform and the distorted BIM [72, 75]. Reconstructions using measurements of clay balls in water at 1.74 GHz are presented in [75], and results from a layered media in [76]. Progress on extension to a 3-D antenna configuration is described in [67].

The University of Michigan system consists of 12 panels in a circular arrangement forming cavity with a 15 cm diameter. Each panel has 3 bow-tie antennas, and designs both for 1.64 GHz and 915 MHz have been used [28]. Data are acquired with a VNA and a switching network and images are formed by using a BIM. Apart from experimental work on acrylic spheres, the system has also been used for hyperthermia experiments at 915 MHz [66].

The Supélec system is an extension of a 2.45 GHz microwave camera used in 2-D mode and for real-time acquisition [19, 34]. Two horn antennas, each having a 22.3 cm by 22.3 cm retina with 32×32 probes, are placed on each side of the rectangular imaging domain, giving a geometry similar to X-ray mammography. However, a (target-)rotator giving multi-view acquisition can also be used. A Newton–Kantorovich algorithm has been used to reconstruct images from measurements in a 2-D geometry (using only one line in the retina), but the system holds potential for 3-D imaging by using the entire retina, and double modulation (use of Tx and Rx retina) could also be applied [29].

6.3   Radar-Based Ultra-Wideband Systems

The systems described in Table 6.3 are at very different maturity levels, ranging from multi-antenna system used for a variety of phantoms to in-air (i.e. without coupling medium) two-antenna experiments.

Despite the lack of published results, attention should also be pointed to a prototype breast imaging system under development at the National University of Ireland Galway. This CMI-based prototype involves a single rotating antenna which scans the breast. The unique characteristic of this prototype is that the entire measurement system moves with the antenna (through a series of computer-controlled stepper and rotational motors), eliminating any artefacts from cable movement in the resultant breast image. The entire measurement system is housed within a portable patient examination table, making the system suitable for early patient trials.

The University of Queensland system has been used with different in-air antenna configurations, e.g. a circular array of 12 UWB antennas for 3–12 GHz. A two-port VNA combined with two Single-Pole 6 Through (SP6T)-switches records data in the frequency domain, which is then transformed to the time-domain. Considered targets include: conductor, dielectric and water cylinders submerged in a larger cylindrical plastic container with canola oil [7, 8]. Recent antenna developments are described in [47].

The University of Manitoba also investigated a pre-clinical UWB prototype. It uses an Anritsu MS2026A VNA (or a Field Fox N9923A from Agilent

Table 6.3 Selected radar-based ultra-wideband imaging systems—apart from the operational University of Bristol system described in Sect. 6.1

Origin	Imaging domain	Antenna configuration	Targets
University of Queensland	Free space	12 Vivaldi antennas Circular array	Canola oil cylinder w. misc. rods
University of Manitoba	Rectangular tub (w. coupling liquid)	1 Vivaldi antenna	Misc. dielectric cylinders
Politecnico di Torino	Cylindrical tank (w. coupling liquid)	8 monopole antennas Circular array	Metal cylinder in glycerin/water
Technical Univ. of Catalonia	Free space (w. absorbers)	1 fixed 1 on linear positioner	Misc. rods and cylinders Clay balls in paraffin
Universitat Rovira i Virgili	Free space (w. absorbers)	1 or 2 fixed monopoles Rotating object	Cylindrical water tank w. dielectric rod
Toyohashi Univ. of Technology	Rectangular tub (w. coupling liquid)	2 Vivaldi antennas Mechanical scanning	Metallic ball in hardened oil

Technologies) to measure S_{11}, from a Vivaldi antenna in a 5.99 GHz band with a centre frequency of 3.01 GHz, [18]. The antenna is attached to a wall in the interior of a plexiglass tank filled with canola oil. Datasets are collected along a circular scan geometry with a 8.25 cm radius, by rotating the phantom. A wide range of phantoms for the cylindrical dielectric targets are described in [18]. One idea is to use the regional tissue information derived from the UWB-radar technique as prior information in an tomographic system, to improve the quantitative dielectric images in comparison with using tomography alone [6]. However, development is on-going to improving both methods individually and in a complementary mode.

Very recently Politecnico di Torino reported on the development of an 8 monopole 2-D system, [69]. It operates in the 0.5–3.0 GHz band and uses the I-MUSIC algorithm for inversion. Results from measurements on a metallic cylinder were reported in [70], and at MiMed COST meetings (TD1301).

The University of Catalunya system consists of two antennas in air. Data are acquired using a VNA. 2-D results from long dielectric cylinders are presented in [24, 25], where two rotary stages were used to allow for imaging of non-symmetric objects. The system and imaging algorithm were extended to allow for imaging of a rotary symmetric 3-D geometry and results from imaging of a short conducting cylinder are given in [26] and from clay balls in paraffin in [27].

The Universitat Rovira i Virgili system consists of one or two monopoles in air illuminating a cylindrical water-filled tank, which can be rotated around its axis of symmetry. Results from 0.5–3 GHz VNA measurements on a low permittivity rod in the tank are provided in [39] and results from an oil-filled tank, in a two antenna configuration using a 5 GHz pulse generator and a 3.1–10.6 GHz receive antenna in [40].

The experiments at Toyohashi University of Technology have been conducted with two mechanically scanned Vivaldi antennas, both in-air and in a tank with matching liquid. The 8 V, 65 ps (FWHM) transmitted pulse is received using a high-speed sampling oscilloscope [79]. A simple background subtraction method has been used on measurements on a breast-shaped phantom made of hardened cooking oil, with an embedded metallic ball with 9 mm diameter.

Also worth mentioning are the measurements conducted at Nanyang Technological University on realistic breast phantoms [38]. An 80 ps (FWHM) Gaussian pulse is generated and received using a 40 GHz sampling oscilloscope and two antennas in a bistatic configuration.

Finally, we would like to point to upcoming results from 2-D experiments by the Second University of Naples, Italy, in collaboration with the Dublin Institute of Technology, Ireland, which were recently presented at PIERS 2015.

References

[1] Aguilar SM, Al-Joumayly MA, Burfeindt MJ, Behdad N, Hagness SC (2014) Multiband Miniaturized Patch Antennas for a Compact, Shielded Microwave Breast Imaging Array. IEEE Transactions on Antennas and Propagation 62(3):1221–1231, DOI 10.1109/TAP.2013.2295615

[2] Al-Joumayly MA, Aguilar SM, Behdad N, Hagness SC (2010) Dual-Band Miniaturized Patch Antennas for Microwave Breast Imaging. IEEE Antennas and Wireless Propagation Letters 9:268–271, DOI 10.1109/LAWP.2010.2045871

[3] Amineh RK, Ravan M, Trehan A, Nikolova NK (2011) Near-field microwave imaging based on aperture raster scanning with TEM horn antennas. IEEE Transactions on Antennas and Propagation 59(3):928–940

[4] Amineh RK, Khalatpour A, Xu H, Baskharoun Y, Nikolova (2012) Three-Dimensional Near-Field Microwave Holography for Tissue Imaging. International Journal of Biomedical Imaging 2012:1–11

[5] Asefi M, OstadRahimi M, Zakaria A, LoVetri J (2014) A 3-d dual-polarized near-field microwave imaging system. Microwave Theory and Techniques, IEEE Transactions on 62(8):1790–1797

[6] Baran A, Kurrant D, Zakaria A, Fear E, LoVetri J (2014) Breast imaging using microwave tomography with radar-based tissue-regions estimation. In: Progress In Electromagnetics Research (PIER), vol 149, pp 161–171

[7] Bialkowski ME, Wang Y, Bakar AA, Khor WC (2010) UWB microwave imaging system including circular array antenna. In: Proceedings on 18th International Conference on Microwave Radar and Wireless Communications (MIKON), 2010, pp 1–4

[8] Bialkowski ME, Wang Y, Bakar AA, Khor WC (2012) Microwave imaging using ultra wideband frequency-domain data. Microwave and Optical Technology Letters 54(1):13–18, DOI 10.1002/mop.26465

[9] Bourqui J, Okoniewski M, Fear EC (2010) Balanced Antipodal Vivaldi Antenna With Dielectric Director for Near-Field Microwave Imaging. IEEE Transactions on Antennas and Propagation 58(7):2318–2326, DOI 10.1109/TAP.2010.2048844

[10] Bourqui J, Garrett J, Fear E (2012a) Measurement and analysis of microwave frequency signals transmitted through the breast. Journal of Biomedical Imaging 2012:1–11

[11] Bourqui J, Sill JM, Fear EC (2012b) A Prototype System for Measuring Microwave Frequency Reflections from the Breast. International Journal of Biomedical Imaging 2012:1–12, DOI 10.1155/2012/851234

[12] Ciocan R, Jiang H (2004) Model-based microwave image reconstruction: simulations and experiments. Medical Physics 31(12):3231–3241

[13] Curtis C, Frayne R, Fear E (2012) Using X-Ray Mammograms to Assist in Microwave Breast Image Interpretation. International Journal of Biomedical Imaging 2012:1–11, DOI 10.1155/2012/235380

[14] Epstein NR, Meaney PM, Paulsen KD (2014) 3d parallel-detection microwave tomography for clinical breast imaging. Review of Scientific Instruments 85(12):124,704, DOI 10.1063/1.4901936

[15] Fear EC, Bourqui J, Curtis C, Mew D, Docktor B, Romano C (2013) Microwave Breast Imaging With a Monostatic Radar-Based System: A Study of Application to Patients. IEEE Transactions on Microwave Theory and Techniques 61(5):2119–2128, DOI 10.1109/TMTT.2013.2255884

[16] Fhager A, Koster J, Rubaek T, Persson M (2011) Modeling and reconstruction in a 3d microwave imaging system. In: General Assembly and Scientific Symposium, 2011 XXXth URSI, pp 1–4

[17] Fhager A, Gustafsson M, Nordebo S (2012) Image Reconstruction in Microwave Tomography Using a Dielectric Debye Model. IEEE Transactions on Biomedical Engineering 59(1):156–166, DOI 10.1109/TBME.2011.2168606

[18] Flores-Tapia D, Pistorius S (2011) Real-time breast microwave radar image reconstruction using circular holography: A study of experimental feasibility. Medical Physics 38(10):5420–5431

[19] Franchois A, Joisel A, Pichot C, Bolomey JC (1998) Quantitative microwave imaging with a 2.45-GHz planar microwave camera. IEEE Transactions on Medical Imaging 17(4):550–561

[20] Gilmore C, Mojabi P, Zakaria A, Ostadrahimi M, Kaye C, Noghanian S, Shafai L, Pistorius S, LoVetri J (2010a) A Wideband Microwave Tomography System With a Novel Frequency Selection Procedure. IEEE Transactions on Biomedical Engineering 57(4):894–904, DOI 10.1109/TBME.2009.2036372

[21] Gilmore C, Mojabi P, Zakaria A, Pistorius S, LoVetri J (2010b) On Super-Resolution With an Experimental Microwave Tomography System. IEEE Antennas and Wireless Propagation Letters 9:393–396, DOI 10.1109/LAWP.2010.2049471

[22] Gilmore C, Zakaria A, Mojabi P, Ostadrahimi M, Pistorius S, Lo Vetri J (2011) The University of Manitoba Microwave Imaging Repository: A two-dimensional microwave scattering database for testing inversion and calibration algorithms. IEEE Antennas and Propagation Magazine 53(5):126–133

[23] Grzegorczyk TM, Meaney PM, Kaufman PA, diFlorio Alexander RM, Paulsen KD (2012) Fast 3-D Tomographic Microwave Imaging for Breast Cancer Detection. IEEE Transactions on Medical Imaging 31(8):1584–1592, DOI 10.1109/TMI.2012.2197218

[24] Guardiola M, Capdevila S, Blanch S, Romeu J, Jofre L (2009a) UWB high-contrast robust tomographic imaging for medical applications. In: Electromagnetics in Advanced Applications, 2009. ICEAA'09. International Conference on, pp 560–563

[25] Guardiola M, Capdevila S, Jofre L (2009b) UWB BiFocusing tomography for breast tumor detection. In: Proceedings of 3rd European Conference on Antennas and Propagation 2009, pp 1855–1859

[26] Guardiola M, Jofre L, Capdevila S, Blanch S, Romeu J (2010) Toward 3d UWB tomographic imaging system for breast tumor detection. In: Proceedings of the Fourth European Conference on Antennas and Propagation (EuCAP), 2010, pp 1–5

[27] Guardiola M, Jofre L, Capdevila S, Blanch S, Romeu J (2011) 3d UWB magnitude-combined tomographic imaging for biomedical applications. Algorithm validation. Radio Engineering 20(2):366–372

[28] Haynes M, Stang J, Moghaddam M (2012) Microwave Breast Imaging System Prototype with Integrated Numerical Characterization. International Journal of Biomedical Imaging 2012:1–18, DOI 10.1155/2012/706365

[29] Henriksson T, Joachimowicz N, Conessa C, Bolomey JC (2010) Quantitative Microwave Imaging for Breast Cancer Detection Using a Planar 2.45 GHz System. IEEE Transactions on Instrumentation and Measurement 59(10):2691–2699, DOI 10.1109/TIM.2010.2045540

[30] Henriksson T, Klemm M, Gibbins D, Leendertz J, Horseman T, Preece AW, Benjamin R, Craddock IJ (2011) Clinical trials of a multistatic UWB radar for breast imaging. In: Loughborough Antennas and Propagation Conference (LAPC), 2011, pp 1–4

[31] Jensen PD, Rubaek T, Mohr JJ, Zhurbenko V (2011) Nonlinear 3-D microwave imaging for breast-cancer screening: Log, phase and log-phase formulation. In: Loughborough Antennas and Propagation Conference (LAPC), 2011, p 4 pages

[32] Jensen PD, Rubak T, Mohr JJ (2012) Utilization of multiple frequencies in 3d nonlinear microwave imaging. In: 6th European Conference on Antennas and Propagation (EUCAP), 2012, pp 1776–1779

[33] Jiang H, Li C, Pearlstone D, Fajardo LL (2005) Ultrasound-guided microwave imaging of breast cancer: Tissue phantom and pilot clinical experiments. Medical Physics 32(8):2528–2535

[34] Joisel A, Bolomey JC (2000) Rapid microwave imaging of living tissues. Proceedings of the SPIE - The International Society for Optical Engineering 3977:320–330

[35] Klemm M, Craddock IJ, Leendertz JA, Preece A, Benjamin R (2009) Radar-Based Breast Cancer Detection Using a Hemispherical Antenna–Experimental Results. IEEE Transactions on Antennas and Propagation 57(6):1692–1704, DOI 10.1109/TAP.2009.2019856

[36] Klemm M, Leendertz JA, Gibbins D, Craddock IJ, Preece A, Benjamin R (2010) Microwave radar-based differential breast cancer imaging: imaging in homogeneous breast phantoms and low contrast scenarios. IEEE Transactions on Antennas and Propagation 58(7):2337–2344

[37] Kwon KC, Lim YT, Kim CH, Kim N, Park C, Yoo KH, Son SH, Jeon SI (2012) Microwave Tomography Analysis System for Breast Tumor Detection. Journal of Medical Systems 36(3):1757–1767, DOI 10.1007/s10916-010-9635-4

[38] Lai JCY, Soh CB, Gunawan E, Low KS (2011) UWB microwave imaging for breast cancer detection—experiments with heterogeneous breast phantoms. Progress In Electromagnetics Research M 16:19–29

[39] Lazaro A, Girbau D, Villarino R (2009a) Simulated and experimental investigation of microwave imaging using UWB. In: Progress In Electromagnetics Research (PIER), vol 94, pp 263–280

[40] Lazaro A, Girbau D, Villarino R (2009b) Wavelet-based breast tumor localization technique using a UWB radar. In: Progress In Electromagnetics Research (PIER), vol 98, pp 75–95

[41] Lee J, Son S, Kim B, Choi H, Jeon S (2014) Animal Testing using 3d Microwave Tomography system for Breast Cancer Detection. e-Health – For Continuity of Care pp 491–495

[42] Li D, Meaney PM, Paulsen KD (2003) Conformal microwave imaging for breast cancer detection. IEEE Transactions on Microwave Theory and Techniques 51(4):1179–1186

[43] Meaney PM, Fanning MW, Li D, Poplack SP, Paulsen KD (2000) A clinical prototype for active microwave imaging of the breast. IEEE Transactions on Microwave Theory and Techniques 48(11):1841–1853

[44] Meaney PM, Fanning MW, Raynolds T, Fox CI, Fang Q, Kogel CA, Poplack SP, Paulsen KD (2007) Initial Clinical Experience with Microwave Breast Imaging in Women with Normal Mammography. Academic Radiology 14(2):207–218, DOI 10.1016/j.acra.2006.10.016

[45] Meaney PM, Golnabi AH, Epstein NR, Geimer SD, Fanning MW, Weaver JB, Paulsen KD (2013a) Integration of microwave tomography with magnetic resonance for improved breast imaging. Medical Physics 40(10):103,101, DOI 10.1118/1.4820361

[46] Meaney PM, Kaufman PA, Muffly LS, Click M, Poplack SP, Wells WA, Schwartz GN, di Florio-Alexander RM, Tosteson TD, Li Z, others (2013b) Microwave imaging for neoadjuvant chemotherapy monitoring: initial clinical experience. Breast Cancer Res 15(2):1–16

[47] Mohammed BJ, Abbosh AM, Sharpe P (2013) Planar array of corrugated tapered slot antennae for ultrawideband biomedical microwave imaging system. International Journal of RF and Microwave Computer-Aided Engineering 23(1):59–66, DOI 10.1002/mmce.20651

[48] Ostadrahimi M, Mojabi P, Noghanian S, Shafai L, Pistorius S, LoVetri J (2012) A Novel Microwave Tomography System Based on the Scattering Probe Technique. IEEE Transactions on Instrumentation and Measurement 61(2):379–390, DOI 10.1109/TIM.2011.2161931

[49] Ostadrahimi M, Mojabi P, Zakaria A, LoVetri J, Shafai L (2013a) Enhancement of Gauss-Newton Inversion Method for Biological Tissue Imaging. IEEE Transactions on Microwave Theory and Techniques 61(9):3424–3434, DOI 10.1109/TMTT.2013.2273758

[50] Ostadrahimi M, Zakaria A, LoVetri J, Shafai L (2013b) A near-field dual polarized (te–tm) microwave imaging system. Microwave Theory and Techniques, IEEE Transactions on 61(3):1376–1384

[51] Padhi SK, Fhager A, Persson M, Howard J (2008) Measured Antenna Response of a Proposed Microwave Tomography System Using an Efficient 3-D FDTD Model. IEEE Antennas and Wireless Propagation Letters 7:689–692, DOI 10.1109/LAWP.2008.2009888

[52] Pallone MJ, Meaney PM, Paulsen KD (2012) Surface scanning through a cylindrical tank of coupling fluid for clinical microwave breast imaging exams. Medical Physics 39(6):3102–3111

[53] Poplack SP, Tosteson TD, Wells WA, Pogue BW, Meaney PM, Hartov A, Kogel CA, Soho SK, Gibson JJ, Paulsen KD (2007) Electromagnetic Breast Imaging: Results of a Pilot Study in Women with Abnormal Mammograms. Radiology 243(2):350–359

[54] Porter E, Fakhoury J, Oprisor R, Coates M, Popovic M (2010) Improved tissue phantoms for experimental validation of microwave breast cancer detection. In: Antennas and Propagation (EuCAP), 2010 Proceedings of the Fourth European Conference on, IEEE, pp 1–5

[55] Porter E, Kirshin E, Santorelli A, Coates M, Popovic M (2013a) Time-domain multistatic radar system for microwave breast screening. Antennas and Wireless Propagation Letters, IEEE 12:229–232

[56] Porter E, Kirshin E, Santorelli A, Popovic M (2013b) A clinical prototype for microwave breast imaging using time-domain measurements. In: Antennas and Propagation (EuCAP), 2013 7th European Conference on, IEEE, pp 830–832

[57] Porter E, Santorelli A, Popovic M (2014) Time-domain microwave radar applied to breast imaging: Measurement reliability in a clinical setting. In: Progress In Electromagnetics Research (PIER), vol 149, pp 119–132

[58] Porter E, Coates M, Popovic M (2015) An early clinical study of time-domain microwave radar for breast health monitoring. IEEE Transactions on Biomedical Engineering PP(99):(early access), DOI 10.1109/TBME.2015.2465867

[59] Rubaek T, Kim OS, Meincke P (2009) Computational Validation of a 3-D Microwave Imaging System for Breast-Cancer Screening. IEEE Transactions on Antennas and Propagation 57(7):2105–2115, DOI 10.1109/TAP.2009.2021879

[60] Santorelli A, Chudzik M, Kirshin E, Porter E, Lujambio A, Arnedo I, Popovic M, Schwartz JD (2013) Experimental demonstration of pulse shaping for time-domain microwave breast imaging. Progress In Electromagnetics Research 133:309–329

[61] Semenov S, Kellam J, Nair B, Williams T, Quinn M, Sizov Y, Nazarov A, Pavlovsky A (2011a) Microwave tomography of extremities: 2. Functional fused imaging of flow reduction and simulated compartment syndrome. Physics in Medicine and Biology 56(7):2019–2030, DOI 10.1088/0031-9155/56/7/007

[62] Semenov S, Kellam J, Sizov Y, Nazarov A, Williams T, Nair B, Pavlovsky A, Posukh V, Quinn M (2011b) Microwave tomography of extremities: 1. Dedicated 2d system and physiological signatures. Physics in Medicine and Biology 56(7):2005–2017, DOI 10.1088/0031-9155/56/7/007

[63] Semenov SY, Svenson RH, Bulyshev AE, Souvorov AE, Nazarov AG, Sizov YE, Posukh VG, Pavlovsky A, Repin PN, Starostin AN (2002) Three-dimensional microwave tomography: Initial experimental imaging of animals. IEEE Transactions on Biomedical Engineering 49(1):55–63

[64] Semenov SY, Posukh VG, Bulyshev AE, Williams TC, Sizov YE, Repin PN, Souvorov A, Nazarov A (2006) Microwave Tomographic Imaging of the Heart in Intact Swine. Journal of Electromagnetic Waves and Applications 20(7):873–890, DOI 10.1163/156939306776149897

[65] Son SH (2010) Preclinical Prototype Development of a Microwave Tomography System for Breast Cancer Detection. ETRI Journal 32(6):901–910, DOI 10.4218/etrij.10.0109.0626

[66] Stang J, Haynes M, Carson P, Moghaddam M (2012) A Preclinical System Prototype for Focused Microwave Thermal Therapy of the Breast. IEEE Transactions on Biomedical Engineering 59(9):2431–2438, DOI 10.1109/TBME.2012.2199492

[67] Stang JP, Joines WT, Liu QH, Ybarra GA, George RT, Yuan M, Leonhardt I (2009) A tapered microstrip patch antenna array for use in breast cancer screening via 3d active microwave imaging. In: Antennas and Propagation Society International Symposium, 2009. APSURSI'09. IEEE, IEEE, pp 1–4

[68] Tafreshi AK, Karadas M, Top CB, Gencer NG (2014) Data acquisition system for harmonic motion microwave doppler imaging. In: Engineering in Medicine and Biology Society, 2014. 36th Annual International Conference of the IEEE, pp 2873–2876

[69] Tobon JAV, Dassano G, Vipiana F, Casu MR, Vacca M, Pulimeno A, Vecchi G (2015a) Design and modeling of a microwave imaging system for breast cancer detection. In: Proceedings of 9th European Conference on Antennas and Propagation 2015, p 2 pages

[70] Tobon JAV, Vipiana F, Dassano G, Casu MR, Vacca M, Pulimeno A, Solimene R (2015b) Experimental results on the use of the MUSIC algorithm for early breast cancer detection. In: Electromagnetics in Advanced Applications, 2015. ICEAA'15. International Conference on, pp 1084–1085

[71] Top CB, Gencer NG (2014) Harmonic motion microwave doppler imaging: A simulation study using a simple breast model. IEEE Transactions on Medical Imaging 33(2):290–300

[72] Wei B, Simsek E, Yu C, Liu QH (2007) Three-dimensional electromagnetic nonlinear inversion in layered media by a hybrid diagonal tensor approximation: Stabilized biconjugate gradient fast Fourier transform method. Waves in Random and Complex Media 17(2):129–147

[73] Williams TC, Bourqui J, Cameron TR, Okoniewski M, Fear EC (2011) Laser Surface Estimation for Microwave Breast Imaging Systems. IEEE Transactions on Biomedical Engineering 58(5):1193–1199, DOI 10.1109/TBME.2010.2098406

[74] Yang F (2013) Microwave imaging for early stage breast tumor detection and discrimination via complex natural resonances. PhD thesis, Faculty of Engineering & Information Technology, University of Technology, Sydney, Australia

[75] Yu C, Yuan M, Stang J, Bresslour E, George RT, Ybarra GA, Joines WT, Liu QH (2008) Active Microwave Imaging II: 3-D System Prototype and Image Reconstruction From Experimental Data. IEEE Transactions on Microwave Theory and Techniques 56(4):991–1000, DOI 10.1109/TMTT.2008.919661

[76] Yu C, Mengqing Yuan, Zhang Y, Stang J, George R, Ybarra G, Joines W, Qing Huo Liu (2010) Microwave Imaging in Layered Media: 3-D Image Reconstruction From Experimental Data. IEEE Transactions on Antennas and Propagation 58(2):440–448, DOI 10.1109/TAP.2009.2037770

[77] Zakaria A, Jeffrey I, LoVetri J (2013) Full-vectorial parallel finite-element contrast source inversion method. Progress In Electromagnetics Research 142:463–483

[78] Zeng X, Fhagei A, Linner P, Persson M, Zirath H (2011) Experimental Investigation of the Accuracy of an Ultrawideband Time-Domain Microwave-Tomographic System. IEEE Transactions on Instrumentation and Measurement 60(12):3939–3949, DOI 10.1109/TIM.2011.2141250

[79] Zhang D, Mase A (2011) Experimental study on radar-based breast cancer detection using UWB antennas without background subtraction. Biomedical Engineering: Applications, Basis and Communications 23(05):383–391, DOI 10.4015/S1016237211002712

[80] Zhurbenko V, Rubaek T, Krozer V, Meincke P (2010) Design and realisation of a microwave three-dimensional imaging system with application to breast-cancer detection. IET Microwaves, Antennas & Propagation 4(12):2200–2211, DOI 10.1049/iet-map.2010.0106

CPSIA information can be obtained
at www.ICGtesting.com
Printed in the USA
LVHW06s2215190818
587450LV00004B/8/P